Math Mammoth
Grade 3 Answer Keys

for the complete curriculum
(International Series)

Includes answer keys to:

- Worktext part A
- Worktext part B
- Tests
- Cumulative Revisions

By Maria Miller

Contents

Math Mammoth
Grade 3-A
Answer Key
International Version

By Maria Miller

Contents

Chapter 1: Addition and Subtraction

Addition Facts Revision, pp. 16-18

Page 16

1. a. 15, 17, 14, 12   b. 13, 15, 11, 18   c. 13, 11, 14, 16   d. 12, 15, 17, 13

Page 17

2. a. 15, 25, 65   b.18, 38, 78   c. 12, 42, 72

3. a. 33   b. 83   c. 76   d. 63   e. 32   f. 56

4.

Page 18

5. a. Add 20.	b. Add 40.	c. Add 15.	d. Add 25.
20	40	15	25
40	80	30	50
60	120	45	75
80	160	60	100
100	200	75	125
120	240	90	150
140	280	105	175

6. and 7. (See the chapter introduction for game rules.)

8. Path: 8 + 6, 7 + 7, 5 + 9, 12 + 2, 9 + 5 (or 6 + 8, then 9 + 5), 7 + 7

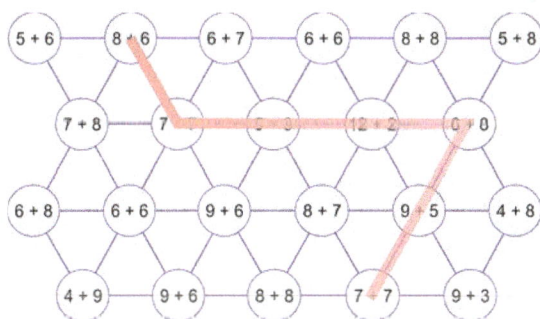

Mental Addition, pp. 19-20

Page 19

1.

a. $50 + 14 = 50 + 10 + 4 = 64$	b. $80 + 11 = 80 + 10 + 1 = 91$	c. $50 + 39 = 50 + 30 + 9 = 89$
d. $35 + 60 = 30 + 5 + 60 = 95$	e. $10 + 5 + 21 = 10 + 5 + 20 + 1 = 36$	f. $29 + 40 + 30 = 20 + 9 + 40 + 30 = 99$

2. a. $50 + 8 = \underline{58}$ b. $80 + 10 = \underline{90}$ c. $80 + 11 = \underline{91}$ d. $70 + 9 = \underline{79}$ e. $60 + 12 = \underline{72}$ f. $100 + 10 = \underline{110}$

Mental Addition, cont.

3. See the chapter introduction.

4. a. 69   b. 90   c. 107   d. 32   e. 89   f. 110

5. a. 68, 69   b. 128, 127   c. 50, 52   d. 236, 235   e. 76, 75   f. 96, 93   g. 98 + 12, 112   h. 62 + 28, 91   i. 53 + 37, 91

Puzzle Corner:

a. $\triangle = 7$ Solution: $\triangle + \triangle + 1 = 15.$ $\triangle + \triangle$ must equal 14. So, $\triangle$ must equal 7.	b. $\hexagon = 3$, $\square = 8$. Solution: $\square + \hexagon = 11$ $\square - \hexagon = 5$ Guess and check is a great strategy here. Take two numbers that add to 11. For example, 6 and 5. Then check their difference (subtract): $6 - 5 = 1$, which does not match, since $\square - \hexagon$ should be 5. So... guess again. Then, we can try next 8 and 3.... which ends up being the correct answer.	c. $\bigcirc = 10$, $\square = 7$ Solution: $\square + \bigcirc = 17$ $\square + \square = 14$ If two rectangles are 14, then one rectangle = 7. Then we tackle the top equation. $7 + \bigcirc = 17$. The circle equals 10.

Revision: Subtraction Facts, pp. 21-22

1.

$13 - 6 = 7$ $13 - 4 = 9$	$13 - 5 = 8$ $13 - 8 = 5$	$13 - 7 = 6$ $13 - 9 = 4$

2.

$14 = \underline{5} + \underline{9}$ $14 = \underline{6} + \underline{8}$ $14 = \underline{7} + \underline{7}$	$15 = \underline{6} + \underline{9}$ $15 = \underline{7} + \underline{8}$	$16 = \underline{7} + \underline{9}$ $16 = \underline{8} + \underline{8}$

3.

$14 - 5 = 9$ $14 - 7 = 7$ $14 - 9 = 5$ $14 - 6 = 8$ $14 - 8 = 6$	$15 - 6 = 9$ $15 - 8 = 7$ $15 - 9 = 6$ $15 - 7 = 8$	$16 - 7 = 9$ $16 - 9 = 7$ $16 - 8 = 8$

4.

11 = _2_ + _9_	12 = _3_ + _9_
11 = _3_ + _8_	12 = _4_ + _8_
11 = _4_ + _7_	12 = _5_ + _7_
11 = _5_ + _6_	12 = _6_ + _6_

5.

11 − 3 = 8	11 − 2 = 9	12 − 4 = 8	12 − 5 = 7
11 − 7 = 4	11 − 5 = 6	12 − 7 = 5	12 − 9 = 3
11 − 9 = 2	11 − 8 = 3	12 − 8 = 4	12 − 7 = 5
11 − 6 = 5	11 − 4 = 7	12 − 6 = 6	12 − 3 = 9

6. $1 + 1 = 2 \rightarrow 2 - 1 = 1$; $2 + 2 = 4 \rightarrow 4 - 2 = 2$; $3 + 3 = 6 \rightarrow 6 - 3 = 3$;

$4 + 4 = 8 \rightarrow 8 - 4 = 4$; $5 + 5 = 10 \rightarrow 10 - 5 = 5$; $6 + 6 = 12 \rightarrow 12 - 6 = 6$;

$7 + 7 = 14 \rightarrow 14 - 7 = 7$; $8 + 8 = 16 \rightarrow 16 - 8 = 8$; $9 + 9 = 18 \rightarrow 18 - 9 = 9$

7. See the end of the book and the chapter introduction.

8. See the chapter introduction.

Puzzle Corner:

a. $23 + 45 + 6 = 74$; or $25 + 43 + 6 = 74$; $26 + 43 + 5 = 74$; $23 + 46 + 5 = 74$; $26 + 45 + 3 = 74$; $25 + 46 + 3 = 74$, and another a similar set of answers where the 2 and 4 (in the tens places) have been swapped.

First, look at which digits add up to 7 in the tens place. If you use 3 and 4 for the tens place for two of the numbers, they add up to 70. Then the rest of the digits would go to the ones places. We would get $2 + 5 + 6 = 13$, and $70 + 13 = 83$, so that won't work.

If instead we use 2 and 4 in the tens places, they add up to 60. Then $3 + 5 + 6 = 14$, and $60 + 14 = 74$. So, the tens digits need to be 2 and 4 (or 4 and 2) and the rest of the digits will go to the ones places in any order.

b. The tens digits for the two 2-digit numbers need to be 2 and 3 (in other words, the two numbers will be 20-something and 30-something). The ones digit can be put into the boxes in any order. For example, these will work:

$24 + 35 + 6 = 65$; $24 + 36 + 5 = 65$; $25 + 34 + 6 = 65$; $25 + 36 + 4 = 65$; $26 + 34 + 5 = 65$; $26 + 35 + 4 = 65$, and another a similar set of answers where the 2 and 3 (in the tens places) have been swapped.

Subtraction Strategies, Part 1, pp. 23-24

1. a. 9, 49   b. 4, 84   c. 9, 29

2. a. 7, 27, 57   b. 4, 34, 74   c. 9, 49, 69   d. 8, 68, 88

Subtraction Strategies, Part 1, cont.

3. a. 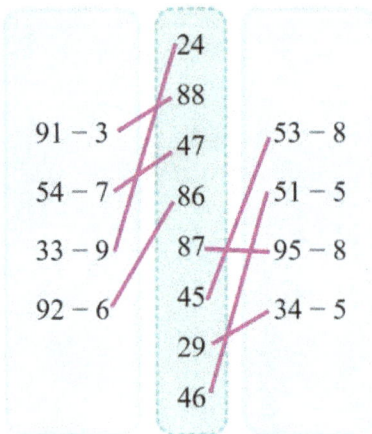   b.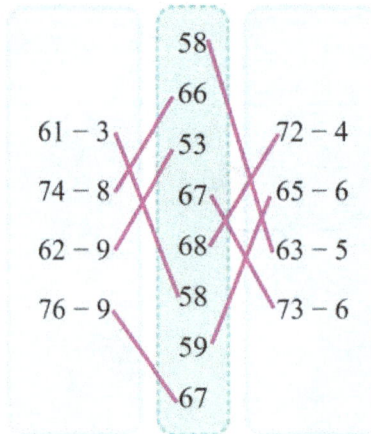

Page 24

4. $50 − $13 − $13 = $24. You have $24 left. Or, $13 + $13 = $26. $50 − $26 = $24.

5. 15 − 7 + 10 = 18 There are 18 children playing on the playground now.

6. 350 + 200 + 150 = 700 m. The lion chased the antelope for 700 metres.

Puzzle Corner:
You will need to subtract 13 from each amount until there is less than 13 left, then count how many times you subtracted 13 to determine how many bouquets of roses you can buy with that amount.

50 − 13 − 13 − 13 = 11. You can buy 3 bouquets of roses with $50.
70 − 13 − 13− 13 − 13 − 13 = 5. You can buy 5 bouquets with $70.
100 − 13 − 13 − 13 − 13 − 13 − 13 − 13 = 9. You can buy 7 bouquets with $100.

Subtraction Strategies, Part 2, pp. 25-26

Page 25

1. a. 64 − 4 − 3 = 57   b. 72 − 2 − 6 = 64   c. 54 − 4 − 4 = 46
   d. 75 − 5 − 2 = 68   e. 27 − 7 − 2 = 18   f. 43 − 3 − 2 = 38

2. See the chapter introduction.

3. a. 89 − 20 − 6 = 63   b. 56 − 30 − 5 = 21   c. 75 − 50 − 1 = 24
   d. 69 − 10 − 9 = 50   e. 67 − 30 − 6 = 31   f. 64 − 30 − 3 = 31

Page 26

4. a. 35; 36   b. 33; 35   c. 34; 35

5. 16; 13   b. 38; 17   c. 48; 43   d. 36; 45

6. No, it is not correct. When subtracting 60 instead of 59, Henry subtracted one too much, so 1 needs *added* to the initial answer of 24. The correct answer is 25.

7. See the chapter introduction.

8. 25; 26; 27   b. 15; 14; 13   c. 47; 46; 45

Page 27

Example 2. 556 + _2_ = 558

1. a. 3, 3   b. 4, 4   c. 6, 6

2.

a. 199 + _15_ = 214 214 − 199 = _15_	b. 67 + _33_ = 100 100 − 67 = _33_

Page 28

3.

a. 92 − 35 = _57_	**b. 805 − 299 = _506_**

35 + _5_ = 40
    40 + _50_ = 90
        90 + _2_ = 92
	299 + _1_ = 300
	    300 + _500_ = 800
	        800 + _5_ = 805

4.

a. 65 − 26 = _39_	b. 83 − 35 = _48_

a. 65 − 26 = _39_

+ 4 | + 30 | + 5
26   30   60   65

b. 83 − 35 = _48_

+ 5 | + 40 | + 3
35   40   80   83

c.	d.	e.	f.
56 − 28 = _28_ 55 − 24 = _31_	72 − 18 = _54_ 82 − 46 = _36_	54 − 37 = _17_ 91 − 57 = _34_	74 − 55 = _19_ 63 − 34 = _29_

5. a. 13 degrees   b. $38

Page 29

4. How can you get a 🦒 into a refrigerator?

Number	0	7	1	6		10	5	1		4	0	0	8
Letter	O	P	E	N		T	H	E		D	O	O	R |,

Number	7	3	10		10	5	1		🦒		2	6
Letter	P	U	T		T	H	E				I	N |,

Number	9	5	3	10		10	5	1		4	0	0	8
Letter	S	H	U	T		T	H	E		D	O	O	R

Mental Maths with Three-Digit Numbers, pp. 30-31

Page 30

1. a. 51, 151   b. 74, 674   c. 23, 323

2. a. 239, 484   b. 662, 522   c. 975, 300

3. See the chapter introduction.

4. a. 259, 89, 239   b. 758, 738, 378   c. 304, 594, 574
   d. 675, 500   e. 680, 888   f. 923, 201

Page 31

5.

196 + 2 = 198	293 + 5 = 298
196 + 3 = 199	293 + 6 = 299
196 + 4 = 200	293 + 7 = 300
196 + 5 = 201	293 + 8 = 301
196 + 6 = 202	293 + 9 = 302
196 + 7 = 203	293 + 10 = 303

6.

a.	b.	c.
393 + 8 = <u>401</u>	797 + 6 = 803	294 + 6 = 300
498 + 5 = 503	993 + 7 = 1,000	497 + 7 = 504
292 + 6 = 298	595 + 8 = 603	291 + 6 = 297

7.

Animal										
Number	4	8	7	2	6	2	5	3	9	9
Letter	Y	O	U	D	I	D	W	E	L	L

A Letter for the Unknown 1, pp. 32-33

Page 32

1. a. 129 + 7 = A; A = 136. Anna is 136 cm tall.
   b. 132 − 5 = J OR J + 5 = 132. J = 127. Jack was 127 cm tall.

2. a. 41 + x = 56 or x + 41 = 56 or x = 56 − 41 or 56 − 41 = x. Solution: x = 15. Ann needs 15 more pins.
   b. 62 + x = 96 or x + 62 = 96 or x = 96 − 62 or 96 − 62 = x. Solution: x = 34. There are 34 pages left to read.

Page 33

Example 3. R = 35

3. Answers will vary because there are several ways to write the equation and also because the student may choose
   any letter for the unknown.

   a. 13 + 18 + c = 50 or 50 − 13 − 18 = c. Solution: c = 19. There are 19 cherry-flavoured candies.
   b. S = 14 + 21 + 5. S = 40. She found 40 socks.
   c. B − 8 = 17 or B = 8 + 17. B = 25. She bought 25 bushes.
   d. C = 61 − 17 or C + 17 = 61. C = 44. She has 44 crayons that are in good shape.
   e. 27 + 4 + x = 45 or x = 45 − 27 − 4. x = 14. She needs $14 more.

4.

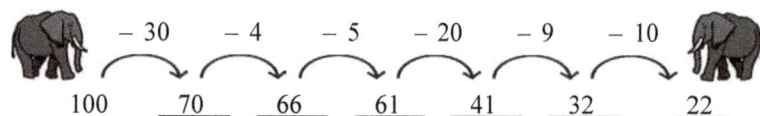

100 70 66 61 41 32 22

The Connection with Addition and Subtraction, pp. 34-35

1.

total 390		total 400	
200 — 190		199 — 201	

a. $200 + 190 = 390$
  $190 + 200 = 390$
  $390 - 200 = 190$
  $390 - 190 = 200$

b. $199 + 201 = 400$
  $201 + 199 = 400$
  $400 - 201 = 199$
  $400 - 199 = 201$

2.

total 95 — 28 — 67

total 99 — 56 — 43

a. $95 - 67 = 28$

b. $99 - 56 = 43$

3.

total 965 — 565 — 400

total 830 — 701 — 129

a. $965 - 400 = 565$

b. $701 + 129 = 830$

Page 35

Teaching box: $x = 370$.

4.

a. $560 + x = 650$ or $x + 560 = 650$
  $x = 90$
	b. $x + 300 = 420$ or $300 + x = 420$
	  $x = 120$

c. total 92 — 37 — x
 	
  $x = 55$
	d. total x — 67 — 52
	
	  $x = 119$

5. a. $210 + x + 200 = 700$; $x = 290$
  b. $x + 28 + 20 = 99$; $x = 51$
  c. $560 + x + 100 = 960$; $x = 300$

Teaching box. Here are some subtraction equations that match the bar model:
(1) $420 - 310 = x$; or turned around, $x = 420 - 310$.
(2) $370 + 50 - 310 = x$ or turned around, $x = 370 + 50 - 310$.

1. a. $x = 1 + 125 + 6$
     $x = 132$. Ernie travelled 132 km.

   b. Students are likely to write two separate equations for this one.
      $46 + 15 = 61$. Suzanne now has 61 points. $72 - 61 = x$. Solution: $x = 11$. He has 11 more points.
      As one equation, it can be written as $x = 72 - 46 - 15$ or as $x = 72 - (46 + 15)$ or $46 + 15 + x = 72$.

   c. $x = 52 - 5 - 9$. $x = 38$. He has 38 small balloons.

2. a. $x = 120 + 52 - 70$. Solution: $x = 102$. She has \$102 left.
   b. Students are likely to write two separate equations for this one. For example: $83 - 11 = 72$ and $x = 72 - 45$.
      As one equation, it can be written as $x = 83 - 11 - 45$. Solution: $x = 27$. Jack has 27 more tennis balls than Rob.
   c. $37 + 15 + x = 66$ or $x = 66 - 37 - 15$. Two equations could be $37 + 15 = 52; x = 66 - 52$.
      Solution: $x = 14$. He still needs \$14.

3.

a. $298 +$ 27 $= 325$	b. $29 +$ 43 $= 72$
325 $-$ 298 $=$ 27	72 $-$ 29 $=$ 43

Puzzle corner.

a. △ $= 22$ ▨ $= 8$ b. ★ $= 21$ c. ⬡ $= 3$ ⬠ $= 10$

Patterns, pp. 38-39

1.

+	5	6	7	8	9	10	11	12	13
5	10	11	12	13	14	15	16	17	18
6	11	12	13	14	15	16	17	18	19
7	12	13	14	15	16	17	18	19	20
8	13	14	15	16	17	18	19	20	21

Answers will vary. Check the student's answers. Here are some patterns students might notice:

- In the columns, the numbers go by 1s. (Each number down is one more than the previous number.)
  This is because each time, we're adding two numbers. When you move down a column, one of the numbers you add does not change (the number at the top of the column) but the other number does, and it is just 1 more than the number that was added previously. For example, $5 + 6$ is just one more than $5 + 5$.

  This principle comes from the associative property of addition, which says that $(a + b) + c = a + (b + c)$.
  Students don't need to know it by name, but they probably realise how it works by now.

  When we move from, say, $6 + 8$ to $6 + 9$, now $6 + 9$ is $6 + (8 + 1)$, which equals $(6 + 8) + 1$. So, $6 + 9$ is just 1 more than $6 + 8$.

Patterns, cont.

1. - In the rows, the numbers go by 1s. (Each number to the right is one more than the previous number.) The explanation is the same as for the previous pattern (the numbers in columns go by 1s).

   - The numbers in the diagonals (from the bottom left towards the upper right) are the same. For example, $8 + 5 = 7 + 6$. In moving along a diagonal, we're changing one addend to be one less and the other to be one more, so the sum stays the same.

   - The numbers in the other diagonals skip-count by twos. This is because in this movement, both addends change by 1, and so, the sum changes by 2.

   - Every other number is even and every other number is odd, both in the rows and in the columns. This is because the numbers count by 1s, and the numbers in the counting sequence alternate between even and odd.

Page 39

2. a. 23, 28, 33, 38, 43. Each time, add 5.
   b. 76, 70, 64, 58, 52. Each time, subtract 6.
   c. 35, 47, 61, 77, 95. Each time, add 2 more than you added the last time. In other words, the differences between the numbers are the even numbers: 2, 4, 6, 8, 10, 12, etc.
   d. 98, 85, 72, 59, 46. Each time, subtract 13.
   e. 156, 216, 286, 366, 456. Each time, add 10 more than you added the last time. The differences between the numbers are multiples of ten: 10, 20, 30, 40, 50, etc.
   f. 350, 410, 470, 530, 590. Each time, add 60.

3. a. See the pattern on the right.

   b. The answers go down by 2s. In other words, each answer is two less than the previous.

   This is because we're subtracting from the same number (340), but each time we're subtracting two more than the previous time.

$340 - 8 = \underline{332}$	
$340 - 10 = \underline{330}$	
$340 - 12 = \underline{328}$	
$340 - 14 = \underline{326}$	
$340 - 16 = \underline{324}$	
$340 - 18 = \underline{322}$	
$340 - 20 = \underline{320}$	
$340 - 22 = \underline{318}$	

4. Going to the right, each number is always 2 more than the previous number in the row. Going down, also, each number is always 2 more than the number above it.

+	34	36	38	40	42
21	55	57	59	61	63
23	57	59	61	63	65
25	59	61	63	65	67
27	61	63	65	67	69

Revision Chapter 1, pp. 40-41

Page 40

1. a. 101   b. 89   c. 103

2. a. 47, 28   b. 74, 33   c. 58, 85

3. a. 308, 304   b. 230, 465   c. 994, 198

4. a. $80 + \underline{240} = 320$; $320 - 80 = 240$ or $320 - 240 = 80$
   b. $900 - 410 = \underline{490}$; $410 + 490 = 900$; $900 - 490 = 410$

17

Page 40

5.

a. $71 - 26 = \underline{45}$ $+$ [4] $+$ [40] $+$ [1] 26 30 70 71	b. $82 - 47 = 35$ c. $63 - 27 = 36$ d. $82 - 51 = 31$ e. $94 - 35 = 59$ f. $45 - 28 = 17$

Page 41

6. a. 3, 5   b. 3, 4   c. 8, 7

7. Student equations will vary; check the student's equations. For example:
   a. $N = 21 + 17 - 5$. $N = 33$. The squirrel has 33 nuts now.
   b. $15 + 5 + M = 28$. $M = 8$. He still needs $8.

8. The numbers in the list are skip-counting by 7. When you add 7 repeatedly, every other number is even and every other number is odd. Since 32, the last number in the list, is even, the next number will be <u>odd</u>.

9. a. 6   b. 7   c. 6

Chapter 2: Regrouping and Rounding

Rounding to the Nearest Ten, Part 1, pp. 47-48

Page 47

Teaching Box
Which one of those is 62 closest to? __60__ And 66? __70__ How about 77? __80__

1.

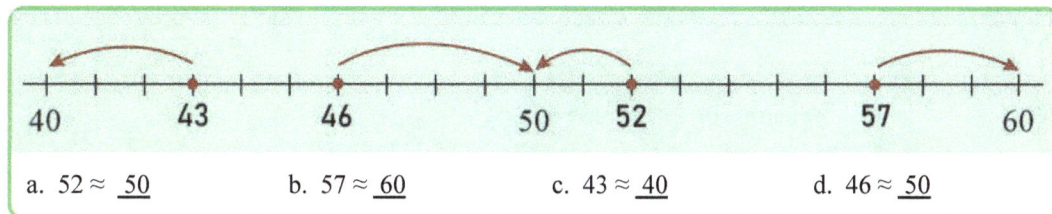

a. 52 ≈ __50__   b. 57 ≈ __60__   c. 43 ≈ __40__   d. 46 ≈ __50__

2. a. 30   b. 50   c. 60   d. 90
   e. 10   f. 30   g. 70   h. 90

3. a. 10   b. 0   c. 10   d. 20   e. 0

Page 48

4. a. 40   b. 70   c. 100   d. 80
   e. 10   f. 70   g. 80   h. 40

5. See the chapter introduction.

6.

a. a skirt, $28, and pants, $33	b. a bicycle, $56, and light, $12	c. a puzzle, $17, and book, $9
a skirt about $30	bicycle about $60	puzzle about $20
pants about $30	light about $10	book about $10
together about $60	together about $70	together about $30

7. Together, they have about 50 + 20 ≈ __70 sacks of apples__.

8. They would cost about $20 + $30 + $30 ≈ __$80__ together.

Rounding to the Nearest Ten, Part 2, pp. 49-51

Page 49

1. a. 240   b. 290   c. 250   d. 300
   e. 270   f. 210   g. 260   h. 300
   i. 310   j. 300   k. 280   l. 240

Page 50

2.

a. 470, 472, 480	b. 820, 829, 830	c. 510, 514, 520
472 ≈ 470	829 ≈ 830	514 ≈ 510
d. 310, 317, 320	e. 600, 608, 610	f. 450, 455, 460
317 ≈ 320	608 ≈ 610	455 ≈ 460

3. a. 600, 890   b. 400, 390   c. 800, 810   d. 110, 1,000

4. a. $38 ≈ $40
   b. 2 weeks ≈ $80; 3 weeks ≈ $120. It will take 4 weeks to earn enough to buy a bicycle.

Page 51

5.

a. a computer, $296, and desk, $188
   computer about $300
   desk about $190
   total cost about $490
	b. a tennis racket, $123, and balls, $38
	   racket about $120
	   balls about $40
	   total cost about $160

c. total cost about $160	d. total cost about $200

6. Across: Down:
   a. 633   a. 655
   b. 796   b. 819
   c. 447   c. 397
   d. 306   d. 512
               e. 911

a. 6	3	0			
6				c. 4	
0		b. 8	0	0	
		2		0	
c. 4	d.5	0			e. 9
	1				1
	0		d. 3	1	0

Regrouping in Addition, pp. 52-54

Page 52

1. a. 231   b. 421   c. 532

Page 53

2. a. 733   b. 642   c. 722   d. 845

3. a. 560   b. 913   c. 859   d. 748

4. a. 772   b. 533   c. 629

Page 54

5. a. 365 + 78 = 443 and 443 + 443 =.886.   Two of the more expensive computers cost $886.
   b. 300 + 300 + 300 – 12 = 888   There are now 888 candles in total.
   c. 285 – 125 = 160   They still have to drive 160 kilometres.

Puzzle corner:

```
    3 1 9          2 8 6
 +  1 9 1       +  6 4 5
   ------         ------
    5 1 0          9 3 1
```

How to Check Addition Problems, pp. 55-57

Page 55

1. The extra ten that was regrouped from the ones column and the extra hundred that was regrouped from the tens column were not added into the totals. The correct answer is 789.

2. a. 1045   b. 938   c. 831   d. 925

How to Check Addition Problems, cont.

3. a. $70 + 90 + 40 + 30 = \underline{\ 230\ }$ b. $270 + 590 + 50 = \underline{\ 910\ }$
   c. $40 + $40 + $90 + $80 = \underline{$250}$ d. $500 + $50 + $30 = \underline{$580}$

4. a. The equations will vary since the student can use any letter for the unknown. These are the main ones we're
      looking for as an answer: $x = 227 + 121 + 266$ or $227 + 121 + 266 = x$
   b. Estimate: 270 ft + 230 ft + 120 ft = $\underline{\ 620\ ft\ }$
   c. The distance all the way around is $\underline{\ 614\ }$ ft. This is reasonable because it is close to our estimate.

5. Equation: $p = 192 + 192 + 48$ or $192 + 192 + 48 = p$
   Solution: 432 pieces.
   (Estimate: $200 + 200 + 50 = 250$ pieces, or $190 + 190 + 50 = 430$ pieces)
   The final answer is reasonable since it is close to the estimate.

6. a. 502   b. 966   c. 399

7. Student equations will vary. For example: $35 + 19 + 22 = d$. Solution: $d = 76$. Dad drove 76 km.
   (Estimate: $40 + 20 + 20 = 80$. The answer of 76 km is reasonable since it's close to the estimate.)

Puzzle Corner:

	3	3	3			5	5	9			4	8	8			3	1	7
+	2	8	7		+	1	7	2		+	4	6	6		+	4	7	8
	6	2	0			7	3	1			9	5	4			7	9	5

Revision: Regrouping in Subtraction, pp. 58-60

1.

a. 4 tens 6 ones → __3__ tens __16__ ones	b. 8 tens 2 ones → __7__ tens __12__ ones
c. 7 tens 4 ones → __6__ tens __14__ ones	d. 6 tens 1 one → __5__ tens __11__ ones

2. a. 36   Check: $36 + 56 = 92$ b. 17   Check: $17 + 38 = 55$ c. 137   Check: $137 + 217 = 354$
   d. 606   Check: $606 + 156 = 762$ e. 239   Check: $239 + 341 = 580$

3. a. Equations: $327 + 50 = 377$ and $495 - 377 = 118$, or $495 - 327 - 50 = 118$. He still needs to save $118.
   b. $7 + 8 = 15$ and $185 - 15 = 170$, or $185 - 7 - 8 = 170$. There were 170 people who did come.
   c. $50 + 50 + 35 + 35 + 25 = 100 + 70 + 25 = 195$. The total weight is 195 kg.

4. a. 272   b. 465   c. 581   d. 61
The number queue: 9 1 6 **5 8 1** 8 3 8 6 7 7 **6 1** 8 8 1 **2 7 2** 6 9 5 8 8 3 6 8 6 6 7 5 **4 6 5** 3 6 6

5. In the ones column, one take away one is not 9 — it is zero.
   He did not need to regroup (borrow) to be able
   to subtract one from one.

The correct way:

```
  8 6 1
- 4 2 1
-------
  4 4 0
```

6. a. 172   Check: $172 + 357 = 529$ b. 461   Check: $461 + 394 = 855$
   c. 380   Check: $380 + 226 = 606$ d. 96   Check: $96 + 541 = 637$

Page 61

Example 1. What is left? _156_

1.

a. 200 + 20 + 1	→ break a 10	200 + 10 + 11 → break a 100

100 + 11 tens + 11

Cross out 97 = _124_.

b. 300 + 40 + 2 → break a 10 → 300 + 30 + 12 → break a 100

200 + 13 tens + 12

Cross out 175 = _167_.

Page 62

1. cont.

c. 300 + 50 → break a 10 → 300 + 40 + 10 → break a 100

200 + 14 tens + 10

Cross out 287 = _63_

d. 400 + 20 + 3 → break a 10 → 400 + 10 + 13 → break a 100

300 + 11 tens + 13

Cross out 156 = _267_.

e. 400 + 30 + 6 → break a 10 → 400 + 20 + 16 → break a 100

300 + 12 tens + 16

Cross out 369 = _67_ .

f. 300 + 10 + 2 → break a 10 → 300 + 0 + 12 → break a 100

200 + 10 tens + 12

Cross out 155 = _157_.

Regrouping Twice in Subtraction, cont.

2. a. 146   b. 268

a. $\begin{array}{r} 600 + 20 + 3 \\ - \quad 400 - 70 - 7 \\ \hline \end{array}$	→	$\begin{array}{r} 600 + 10 + 13 \\ - \quad 400 - 70 - 7 \\ \hline \end{array}$	→	$\begin{array}{r} 500 + 110 + 13 \\ - \quad 400 - 70 - 7 \\ \hline 100 + 40 + 6 \end{array}$
b. $\begin{array}{r} 800 + 50 + 2 \\ - \quad 500 - 80 - 4 \\ \hline \end{array}$	→	$\begin{array}{r} 800 + 40 + 12 \\ - \quad 500 - 80 - 4 \\ \hline \end{array}$	→	$\begin{array}{r} 700 + 140 + 12 \\ - \quad 500 - 80 - 4 \\ \hline 200 + 60 + 8 \end{array}$

3. a. 197   b. 68   c. 159

a. $\begin{array}{r} 700 + 40 + 6 \\ - \quad 500 - 40 - 9 \\ \hline \end{array}$	→	$\begin{array}{r} 700 + 30 + 16 \\ - \quad 500 - 40 - 9 \\ \hline \end{array}$	→	$\begin{array}{r} 600 + 130 + 16 \\ - \quad 500 - 40 - 9 \\ \hline 100 + 90 + 7 \end{array}$
b. $\begin{array}{r} 400 + 60 + 1 \\ - \quad 300 - 90 - 3 \\ \hline \end{array}$	→	$\begin{array}{r} 400 + 50 + 11 \\ - \quad 300 - 90 - 3 \\ \hline \end{array}$	→	$\begin{array}{r} 300 + 150 + 11 \\ - \quad 300 - 90 - 3 \\ \hline 60 + 8 \end{array}$
c. $\begin{array}{r} 900 + 10 + 4 \\ - \quad 700 - 50 - 5 \\ \hline \end{array}$	→	$\begin{array}{r} 900 + 0 + 14 \\ - \quad 700 - 50 - 5 \\ \hline \end{array}$	→	$\begin{array}{r} 800 + 100 + 14 \\ - \quad 700 - 50 - 5 \\ \hline 100 + 50 + 9 \end{array}$

4. a. $32 + 32 + 32 = 96$ and $100 - 96 = 4$, or $100 - 32 - 32 - 32 = 4$. Yes, you can buy three backpacks and you will have $4 left over.

   b. $57 - 12 = 45$ and $45 + 45 = 90$.   The total cost is $90.

Regrouping Twice in Subtraction, Part 2, pp. 65-67

1. a. 147   b. 469   c. 95   d. 165
   e. 776   f. 197   g. 386   h. 188
The number queue: 1 9 7 3 9 5 2 9 1 6 5 4 7 8 3 8 6 7 7 6 1 8 8 1 4 7 4 6 9 5 8 8 6 6 7 5 4 5 8

2. a. 655, $655 + 156 = 811$ b. 366, $366 + 277 = 643$
   c. 199, $199 + 266 = 465$ d. 254, $254 + 657 = 911$

3. Student equations and estimations will vary.

   a. $s = 775 - 248 - 195$. Solution: $s = 332$. There are 332 striped shirts.
      Estimation: $780 - 250 - 200 = 330$.

   b. This one can be tricky to write as a single equation, so, it is reasonable to accept the student writing two equations.
      $d = 365 - 178 + 29$ or $178 - 29 + d = 365$. Or, as two equations: $178 - 29 = 149$ and $365 - 149 = 216$.
      Solution: $d = 216$. There were 216 days that it did not rain in 2023.
      Estimation: $180 - 30 = 150$ and $365 - 150 = 215$.

Regrouping Twice in Subtraction, Part 2, cont.

Page 67

4. a. <   b. >   c. <
   d. <   e. >   f. >

5. a. 499   b. 556   c. 179   d. 367
   e. 277   f. 258   g. 166   h. 379
   i. 376   j. 176   k. 577   l. 77

Puzzle corner:

$$
\begin{array}{r}
4\ \ 3\ \ 8 \\
-\ 1\ \ 2\ \ 3 \\
\hline
3\ \ 1\ \ 5
\end{array}
\qquad
\begin{array}{r}
8\ \ 5\ \ 3 \\
-\ 3\ \ 3\ \ 6 \\
\hline
5\ \ 1\ \ 7
\end{array}
\qquad
\begin{array}{r}
6\ \ 1\ \ 9 \\
-\ 3\ \ 5\ \ 5 \\
\hline
2\ \ 6\ \ 4
\end{array}
\qquad
\begin{array}{r}
6\ \ 8\ \ 4 \\
-\ 4\ \ 7\ \ 7 \\
\hline
2\ \ 0\ \ 7
\end{array}
$$

Regrouping with Zero Tens, pp. 68-70

Page 68

Example 1. 126   Example 2. 139

Page 69

1. a. 172   b. 39   c. 166

Regrouping with Zero Tens, cont.

2. a. 176   b. 319   c. 184

a.
$$700 + 0 + 3$$
$$-\ 500 - 20 - 7$$
$$\rightarrow$$
$$600 + \boxed{100} + 3$$
$$-\ 500 - 20 - 7$$
$$\rightarrow$$
$$600 + \boxed{90} + \boxed{13}$$
$$-\ 500 - 20 - 7$$
$$\overline{100 + 70 + 6}$$

b.
$$600 + 0 + 0$$
$$-\ 200 - 80 - 1$$
$$\rightarrow$$
$$500 + \boxed{100} + 0$$
$$-\ 200 - 80 - 1$$
$$\rightarrow$$
$$500 + \boxed{90} + \boxed{10}$$
$$-\ 200 - 80 - 1$$
$$\overline{300 + 10 + 9}$$

c.
$$800 + 0 + 1$$
$$-\ 600 - 10 - 7$$
$$\rightarrow$$
$$700 + \boxed{100} + 1$$
$$-\ 600 - 10 - 7$$
$$\rightarrow$$
$$700 + \boxed{90} + \boxed{11}$$
$$-\ 600 - 10 - 7$$
$$\overline{100 + 80 + 4}$$

3. Both $27 + b = 83$ *and* $83 - b = 27$ match the problem. Solution: there are 56 boys.

4. Answers will vary. Please check the student's work. For example:
   There are 61 horses in the race. Twenty-two of them are white and the rest are brown. How many horses are brown?

   $22\ +\ \triangle = 61$. Answer: 39 are brown.

Regrouping with Zero Tens, Part 2, pp. 71-73

1. a. 437   b. 275   c. 78   d. 167
   e. 789   f. 128   g. 398   h. 158

2. a. 245; 245 + 556 = 801   b. 324; 324 + 279 = 603
   c. 259; 259 + 266 = 525   d. 448; 448 + 452 = 900

3. a. 605 + 128 = 733   Annie earned $733.
   b. 129 − 20 + 109 = 218   The two cameras will cost $218.
   c. 300 − 65 − 125 = 110   There are 110 red sweaters.

4. a. 97   b. 125   c. 115
   d. 41   e. 60   f. 91

5. a. 25   b. 43   c. 29

6. See the chapter introduction.

7. a. 389   b. 269   c. 359   d. 265
   e. 92   f. 726   g. 149   h. 158

Regrouping with Zero Tens, Part 2, cont.

Page 73

Puzzle corner:

```
    6   0   8          8   0   0          6   0   1          6   1   0
-   2   9   3      -   2   3   6      -   3   5   7      -   4   0   3
  ─────────────      ─────────────      ─────────────      ─────────────
    3   1   5          5   6   4          2   4   4          2   0   7
```

Addition, Subtraction, and Brackets, pp. 74-76

Page 74

1. a. 12, 16   b. 28, 28   c. 16, 12

2. a. $120 - (40 + 50) = 30$
   b. $70 + 50 - 90 = 30$ or $(70 + 50) - 90 = 30$
   c. $50 - (10 - 5) = 45$

3. Jess is correct. Mary got her result by first adding $6 + 2$, and then subtracting that from 20. So, she calculated $20 - (6 + 2)$. Jess calculated it correctly: first subtracting $20 - 6 = 14$, and then adding 2 to that.

Page 75

4. a. 0, 10   b. 140, 0

5. Answers will vary. Here are some possibilities:

   $100 - (50 + 20 - 10) = 40$
   $100 - (50 + 20) - 10 = 20$
   $100 - 50 + (20 - 10) = 60$
   $(100 - 50) + (20 - 10) = 60$

6. Henry is not correct. He calculated $25 - 10 - (4 + 6)$. The correct answer is 17. You could say to Henry that additions and subtractions are done from left to right, in order, so, we will first do $25 - 10 = 15$. Then, $15 - 4 = 11$, and lastly $11 + 6 = 17$.

7. a. $505 - 317 = 188 + 195 = 383$ b. $364 + (409 - 238) = 535$

8.

a. $10 - (5 - 2) = 7$	b. $20 - (5 - 2) - 1 = 16$	c. $15 - (5 + 2 - 1) = 9$
d. $10 - (5 + 2) = 3$	e. $20 - (5 - 2 - 1) = 18$	f. $15 - (5 + 2) - 1 = 7$

Word Problems Practice, pp. 76-78

Page 76

Equations will vary. The ones given are just examples.

1. Equation: $p = 519 - 30 - 17$. Estimate: $520 - 30 - 20 = 470$. Solution: $p = 472$. The final price was $472.

Page 77

2. Equation: $T = 21 + 43 + 43 - 10$. Estimate: $20 + 40 + 40 - 10 = 90$. Solution: $T = 97$. He counted 97 trees.

3. Equation: $124 + 11 + x = 152$. Estimate: $120 + 10 + x = 150; x = 20$. Solution: $x = 17$. The second price increase was $17.

4. Equation: $L = 145 - 32 - 32$. Estimate: $150 - 30 - 30 = 90$. Solution: $L = 81$. She will have $81 left.

Word Problems Practice, cont.

Page 78

5. Equation: 29 + S = 64 + 17. Estimate: He has about $30, and wants to buy things that cost about $80, so he needs to save about $50. Solution: S = 64 + 17 − 29 = 52. He needs to save $52.

6. Equation: S = 32 + 15 + 12 + 19. Estimate: There were about 30 + 20 + 10 + 20 = 80 strawberries. Solution: S = 78. There were 78 strawberries to start with.

7. Equation: B = 16 + 21 + 16 + 10. Estimate: 15 + 20 + 15 + 10 = 60. Solution: B = 63. They have 63 blocks together.

Puzzle corner. a. 61 − 45 = 16   b. 54 − 16 = 38.

Distance Table, pp. 79-80

Page 79

1. 155 km

2. 523 km

3. about 290 km

4. 299 km + 299 km = 598 km

5. 305 km + 305 km = 610 km

Page 80

6. 208 km + 208 km + 289 km = 705 km

7. a. 351 − 92 = 259. He still had 259 km to go.
   b. 259 − 34 = 225. He still had 225 km to drive.

8. 527 − 425 = 102. It is 102 km farther.

9. Almost, but not quite. In six hours he would drive
   70 + 70 + 70 + 70 + 70 + 70 = 420 km of the 425-km journey,
   so he would still have 5 km farther to go.

Mixed Revision Chapter 2, pp. 81-82

Page 81

1.

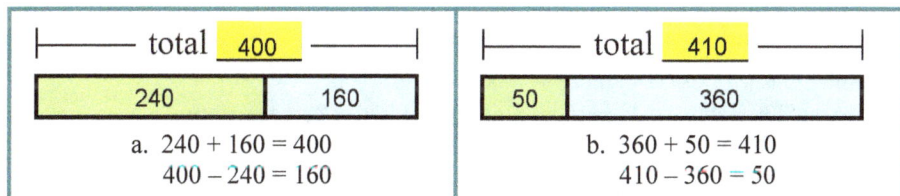

a. 240 + 160 = 400
   400 − 240 = 160

b. 360 + 50 = 410
   410 − 360 = 50

2.

a. 53 − 7 53 − 3 − 4 = __46__	b. 65 − 8 65 − 5 − 3 = __57__	c. 32 − 8 32 − 2 − 6 = __24__
d. 74 − 6 74 − 4 − 2 = __68__	e. 81 − 4 81 − 1 − 3 = __77__	f. 63 − 6 63 − 3 − 3 = __57__

3.

a. 76 − 31 76 − 30 − 1 = __45__	b. 84 − 34 84 − 30 − 4 = __50__	c. 96 − 52 96 − 50 − 2 = __44__
d. 78 − 15 78 − 10 − 5 = __63__	e. 58 − 36 58 − 30 − 6 = 22	f. 85 − 44 85 − 40 − 4 = 41

4. a. 4   b. 3   c. 8

Mixed Revision Chapter 2, cont.

5. a.
$193 + \underline{2} = \underline{195}$
$193 + \underline{4} = \underline{197}$
$193 + \underline{6} = \underline{199}$
$193 + \underline{8} = \underline{201}$
$193 + \underline{10} = \underline{203}$
$193 + \underline{12} = \underline{205}$

   b. The answer increases by two every time, because the number being added is increasing by two.

6. Equations will vary because there are often several ways to write the equation.
   a. $25 - s = 11$; The shirt cost $14.
   b. $60 + 47 + 19 + A = 147$; The clothes for Alex cost $21.
   c. $5 + 9 + m = 21$; The third video was 7 minutes.

Puzzle corner. a. 452   b. 978 (count the letters in each word)

Revision Chapter 2, pp. 83-84

1. a. 50   b. 230   c. 970

2. $40 + 20 + 20 = 80$; His total was about $80.

3. $70 + $70 + $70 + $70 = $280; She could buy the camera in 4 weeks.

4. a. 41   b. 390

5. $100 - 14 = 86$; $100 + 100 + 100 + 86 = 386$; They have 386 straws.

6. a. 139   b. 294   c. 378   d. 377   e. 166

7. a. 27   b. 785

8. a. Student equations will vary. For example: $266 + 258 + Y = 800$ or $Y = 800 - 266 - 258$.
   Estimate: The blue and red beads are about $270 + 260 = 530$ in total. Since $530 + 270 = 800$, there are about 270 yellow beads.
   Solution: $Y = 276$. There are 276 yellow beads. The answer of 276 is reasonable because it's close to the estimate.

   b. Equation: $T = 210 + 178 + 149 + 239$.
   Estimate: $180 + 150 + 210 + 240 = 780$.
   Solution: $T = 776$. There were 776 balls at the start of the month. This is a reasonable answer because it is close to the estimate.

DiffTriangles!

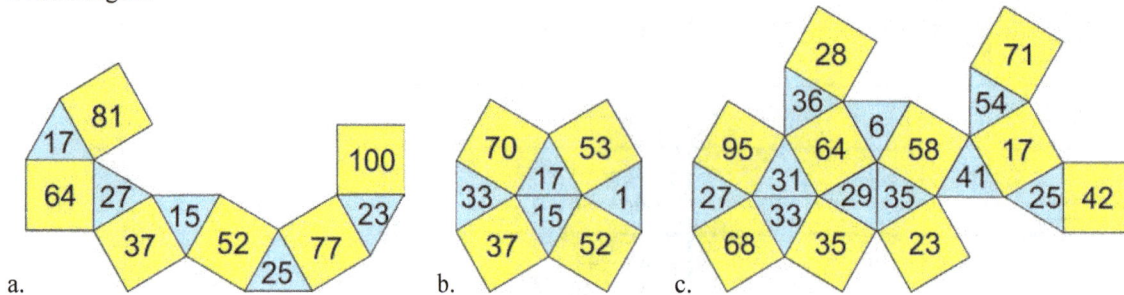

a.   b.   c.

Chapter 3: Concept of Multiplication

Many Times the Same Group, pp. 89-91

1. b. 3×6 c. 5×4
   d. 3×1 e. 5×2 f. 1×7

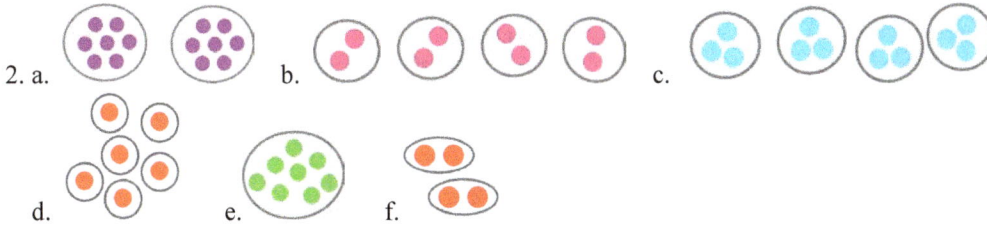

2. a.   b.   c.

   d.   e.   f.

3.

a. $3 + 3 + 3 + 3$ 4 groups of 3 chicks in each. 4×3 chicks $= 12$ chicks	b. $2 + 2 + 2$ 3 groups of 2 hens in each. 3×2 hens $= 6$ hens
c. $1 + 1 + 1$ 3 groups of 1 dog in each. 3×1 dog $= 3$ dogs	d. 1 group of 4 carrots in it. 1×4 carrots $= 4$ carrots
e. $2 + 2 + 2 + 2$ $4 \times 2 = 8$	f. $4 + 4 + 4$ $3 \times 4 = 12$

4.

a. Draw 3 groups of seven sticks. $3 \times 7 = 21$	b. Draw 2 groups of eight circles. $2 \times 8 = 16$
c. Draw 4 groups of one circle. $4 \times 1 = 4$	d. Draw 5 groups of two sticks. $5 \times 2 = 10$

5.

a. $5 \times 4 = 20$	b. $4 \times 6 = 24$

6. a. $5 \times 4 = 20$ b. $6 \times 2 = 12$
   c. $8 \times 2 = 16$ d. $3 \times 11 = 33$

Multiplication as an Array, pp. 92-93

Page 92

1. See Multiplication Arrays under Games & Activities in the chapter introduction.

2.

a. 2 rows, 5 carrots in each row.	b. 3 rows, 3 chicks in each row.
$5 + 5$	$3 + 3 + 3$
$2 \times 5 = 10$ carrots	$3 \times 3 = 9$ chicks
c. 3 rows, 1 bear in each row.	d. 3 rows, 5 bulbs in each row.
$1 + 1 + 1$	$5 + 5 + 5$
$3 \times 1 = 3$ bears	$3 \times 5 = 15$ bulbs

Page 93

3. a. $5 + 5 + 5 = 15$; $3 \times 5 = 15$.   b. $3 + 3 + 3 = 9$; $3 \times 3 = 9$.   c. $10 + 10 + 10 = 30$; $3 \times 10 = 30$.

4. $5 \times 4 = 20$ or $4 \times 5 = 20$

5. He had five columns.

6. Four columns.

Puzzle corner. $96 - 24 = 72$.

Multiplying on a Number Line, pp. 94-96

Page 94

1. a. $7 \times 2 = 14$ b. $4 \times 4 = 16$ c. $5 \times 3 = 15$ d. $7 \times 1 = 7$

Page 95

2.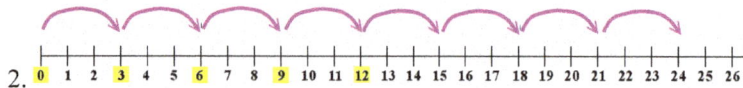

3. a. 15, 12   b. 24, 21   c. 18, 9   d. 6, 27

4. a. 8, 3   b. 6, 5   c. 7, 4   d. 2, 1

5.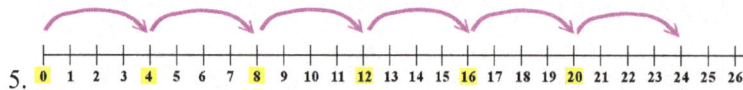

6. a. 8, 16   b. 24, 28   c. 32, 12   d. 20, 4

7. a. 6, 2   b. 0, 3   c. 4, 2   d. 5, 1

Multiplying on a Number Line, cont.

8. a. $6 \times 4 = 24$

b. $5 \times 5 = 25$

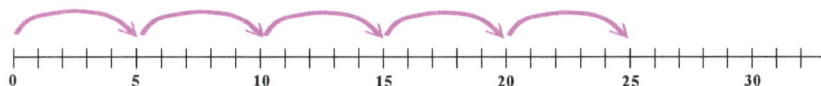

c. $6 \times 5 = 30$

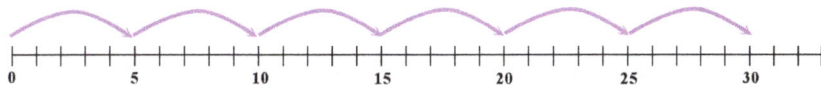

d. $7 \times 4 = 28$

e. $3 \times 10 = 30$

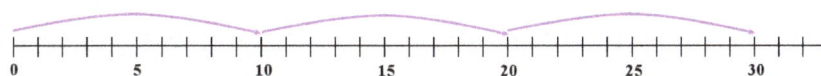

9. a. 6, 18, 20   b. 10, 28, 24   c. 30, 27, 40   d. 30, 22, 21

Order of Operations 1, pp. 97-98

1. a. 19   b. 11   c. 40

2. The part to do first is highlighted.

a. $5 + 4 \times 2$ $5 + 8 = 13$	b. $10 + 5 \times 4$ $10 + 20 = 30$	c. $20 - 4 \times 4$ $20 - 16 = 4$
d. $2 \times 6 + 2 \times 7$ $12 + 14 = 26$	e. $3 \times 5 - 2 \times 4$ $15 - 8 = 7$	f. $2 \times 5 + 1 \times 4$ $10 + 4 = 14$
g. $5 + 1 \times 2 + 5$ $5 + 2 + 5 = 12$	h. $30 - 2 \times 2 - 10$ $30 - 4 - 10 = 16$	i. $50 - 3 \times 2 + 6$ $50 - 6 + 6 = 50$

Teaching Box Example

$$4 \times 2 \;+\; 3 \times 6 \;=\; \underline{\;\;26\;\;}$$
$$\;\;\;8\qquad\quad 18$$

3.

a. $\underline{\;4\;} \times \underline{\;3\;} \;+\; \underline{\;3\;} \times \underline{\;5\;} \;=\; \underline{\;27\;}$
b. $\underline{\;3\;} \times \underline{\;4\;} \;+\; \underline{\;4\;} \times \underline{\;3\;} \;=\; \underline{\;24\;}$
c. $\underline{\;4\;} \times \underline{\;4\;} \;+\; \underline{\;7\;} \;=\; \underline{\;23\;}$
d. $\underline{\;4\;} \times \underline{\;5\;} \;+\; \underline{\;3\;} \times \underline{\;6\;} \;+\; \underline{\;8\;} \;=\; \underline{\;46\;}$
e. $4 \times 10 + 2 \times 6 + 7 = 59$

Understanding Word Problems, Part 1, pp. 99-101

Page 99

1. a. $4 \times 6 = 24$ tennis balls   b. $31 - 26 = 5$ missing balls
    c. $24 - 9 = 15$ left   d. $3 \times 5 = 15$ workers
    e. $9 + 7 = 16$ f. $5 \times 10 = 50$ jumping jacks

Page 100

2. a. $3 \times \$10 + \$20 = \$50$ b. $2 \times \$3 + 3 \times \$4 = \$18$
    c. $\$13 + 5 \times \$5 = \$38$
    d. $3 \times \$2 + 5 \times \$4 = \$26$ e. $4 \times \$3 + 3 \times \$2 = \$18$

3. a. $5 \times \$2 + \$5 = \$15$
    b. $4 \times \$10 + 2 \times \$20 = \$80$

Page 101

4. a. $\$2 + \$2 + 4 \times \$3 = \$16$. The total bill was \$16.
    b. $8 \times 3 = 24$ (or $3 \times 8 = 24$)
    c. $4 \times 5 = 20$. She has flowers in these pots.
    d. $4 \times 4 - 4 = 12$. Her drawing has 12 hearts now.
    e. $4 \times 4 = 16$ or $(3 + 1) \times 4 = 16$. There were 16 pizza slices.
    f. $10 \times 2 + 2 \times 1 = 22$. The bottles weigh 22 kg together.

Puzzle Corner:

    $2 \times 4 + 1 = 9$ $5 + 5 \times 4 = 25$ $5 \times 2 + 5 + 5 = 20$ or $5 + 2 \times 5 + 5 = 20$

Zero and One in Multiplication, pp. 102-103

Page 102

1. a. 0, 0   b. 1, 9   c. 0, 10   d. 6, 0

Page 103

Table of Zero	
$1 \times 0 = \underline{0}$	$7 \times 0 = \underline{0}$
$2 \times 0 = \underline{0}$	$8 \times 0 = \underline{0}$
$3 \times 0 = \underline{0}$	$9 \times 0 = \underline{0}$
$4 \times 0 = \underline{0}$	$10 \times 0 = \underline{0}$
$5 \times 0 = \underline{0}$	$11 \times 0 = \underline{0}$
$6 \times 0 = \underline{0}$	$12 \times 0 = \underline{0}$

Table of One	
$1 \times 1 = \underline{1}$	$7 \times 1 = \underline{7}$
$2 \times 1 = \underline{2}$	$8 \times 1 = \underline{8}$
$3 \times 1 = \underline{3}$	$9 \times 1 = \underline{9}$
$4 \times 1 = \underline{4}$	$10 \times 1 = \underline{10}$
$5 \times 1 = \underline{5}$	$11 \times 1 = \underline{11}$
$6 \times 1 = \underline{6}$	$12 \times 1 = \underline{12}$

3. a. $4 \times 6 - 2 = 22$ There were 22 good eggs.
    b. $3 \times 20 = 60$ Mary has 60 things in the jars.

4. a. 35, 30   b. 1, 0   c. 67, 0

5.

×	0	1	2	3	4
1	0	1	2	3	4
2	0	2	4	6	8
3	0	3	6	9	12
4	0	4	8	12	16

Understanding Word Problems, Part 2, pp. 104-106

Page 104

1. a. $3 \times 6 = 18$ There are six students in each group.
   b. $9 \times \$2 = \$18$ She bought nine notebooks.
   c. $4 \times \$5 = \$20$ He can buy four balls.
   d. $5 \times \$3 = \$15$ The total cost was $15.

Page 105

2. a. $4 + 4 + 4 + 4 - 1 = 15$ or $4 \times 4 - 1 = 15$ The people got 15 pieces of pizza.
   b. $5 + 5 + 5 + 5 + 5 = 25$ or $5 \times 5 = 25$ She will need five boxes.
   c. $6 + 6 + 6 + 6 + 6 = 30$ or $5 \times 6 = 30$ She needs to learn 30 words.
      $2 + 2 + 2 + 2 + 2 = 10$ or $5 \times 2 = 10$ Ten words are in bold.
   d. $2 + 5 + 7 = 14$ He bought 14 pieces of fruit.
   e. $4 + 4 + 4 + 4 + 3 = 19$ or $4 \times 4 + 3 = 19$ There are 19 students in the class.

3. a. 20, 30   b. 0, 78   c. 18, 14

Page 106

4.

a.	b.
$35 \times 1 = 35$	$6 \times 5 = 30$
$1 \times 1 = 1$	$1 \times 0 = 0$
$10 \times 3 = 30$	$67 \times 1 = 67$
c.	d.
$1 \times 45 = 45$	$7 \times 2 = 14$
$0 \times 1 = 0$	$0 \times 0 = 0$
$0 \times 99 = 0$	$0 \times 10 = 0$

5.

×	0	1	2	3	4	5
0	0	0	0	0	0	0
1	0	1	2	3	4	5
2	0	2	4	6	8	10
3	0	3	6	9	12	15
4	0	4	8	12	16	20
5	0	5	10	15	20	25

6. $140 + 140 + 50 = 330$ or $140 \times 2 + 50 = 330$; Her total was about $330.

7. $5 \times 4 + G = 24$; There are 4 green blocks.

8. $12 + 12 + e = 31$ or $2 \times 12 + e = 31$; There are 7 eggs in the third carton.

Puzzle Corner:

$3 \times 8 = 24$ $4 \times 7 = 28$ $5 \times 6 = 30$ $2 \times 9 = 18$

Multiplication in Two Ways, pp. 107-109

Page 107

1.

$4 + 4 + 4 = \underline{12}$ $3 \times 4 = \underline{12}$	$3 + 3 + 3 + 3 = \underline{12}$ $4 \times 3 = \underline{12}$
$2 + 2 + 2 + 2 + 2 = \underline{10}$ $5 \times 2 = \underline{10}$	$\underline{5} + \underline{5}$ $2 \times 5 = \underline{10}$
$1 \times 4 = \underline{4}$	$4 \times 1 = \underline{4}$

What do you notice? Student answers will vary but should indicate the results/answers are the same either way.

Page 108

2. The student's answers may be in either order.

a. $2 \times 3 = 6$	$3 \times 2 = 6$
b. $1 \times 3 = 3$	$3 \times 1 = 3$
c. $3 \times 3 = 9$	$3 \times 3 = 9$

Note: when the number of things in each row is the same as the number of things in each column (when the array is square), the facts are the same both ways.

3. Yes, the pictures work. One of them shows 5 groups of 3 dots, and the other shows 3 groups of 5 dots, both having 15 dots.

4. The student is asked to answer only the multiplication that is easier for them to solve, but many will probably fill in both answers.

a. $2 \times 10 = 20$ OR $10 \times 2 = 20$	b. $7 \times 2 = 14$ OR $2 \times 7 = 14$
Two groups of ten Ten groups of two	Seven groups of two Two groups of seven
c. $3 \times 4 = 12$ OR $4 \times 3 = 12$	d. $11 \times 3 = 33$ OR $3 \times 11 = 33$

Page 109

5. Both of the number line jump pictures show that the jumps end at 20. So, both 5×4 and 4×5 equal 20.

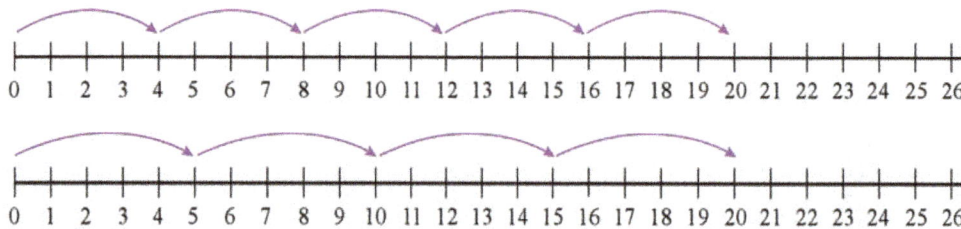

6. a. $7 \times 4 = 28$; $4 \times 7 = 28$
   b. $7 \times 1 = 7$; $1 \times 7 = 7$

7. a. $5 \times 4 = 20$ He has 20 rocks in his buckets.
   b. $7 \times 4 = 28$ There are 7 sheep in the yard.
   c. $3 \times 7 = 21$ The array has seven rows of dots..
   d. $5 \times 4 = 20$ You can make four groups of sticks.

Order of Operations 2, pp. 110-112

Page 110

1. a. Total number of plates = $3 \times (6 + 3) = 27$.
   b. Total number of pieces of fruit = $10 \times (2 + 3) = 50$.
   c. Total number of flowers = $4 \times (5 + 8) = 52$.

2. a. The cost for <u>four</u> shirts is $\underline{4} \times (12 - \underline{2}) = \underline{\$40}$.
   b. $3 \times (10 - 2) = \$24$ OR with $-symbols: $3 \times (\$10 - \$2) = \$24$

Page 111

3. The first step is highlighted.

a. $20 + 6 - 3 = 23$ b. $20 + (6 - 3) = 23$	c. $20 - 6 + 3 = 17$ d. $20 - (6 + 3) = 11$	e. $80 - 30 - (30 + 20) = 0$ f. $80 - (30 - 30) + 20 = 100$

4. The first step is highlighted.

a. $3 + 5 \times 2 = 13$	b. $5 \times (3 + 1) = 20$	c. $4 \times (4 - 2) = 8$
d. $3 \times 6 - 11 = 7$	e. $25 - 5 \times 2 = 15$	f. $(3 - 2) \times 6 = 6$

5. a. $p = 10 \times 2 - 1$. Solution: $p = 19$. There are 19 plates on the table.
   b. $b = 4 \times (2 + 3)$. Solution: $b = 20$. There are 20 babies in total.

Page 112

6. a. $x = 5 \times 2 + 4 \times 4$. Solution: $x = 26$. 26 people can be seated in total.
   b. $t = 5 \times (1 + 1)$ or $t = 5 \times 2$. Solution: $t = 10$. There are 10 towels.

7. The first step is highlighted.

a. $0 \times 7 + 2 = 2$	b. $(10 - 5) \times 4 = 20$	c. $5 \times (1 + 9) = 50$
d. $100 - 5 \times 7 = 65$	e. $8 + 4 \times 3 = 20$	f. $(3 + 2) \times 5 = 25$

8. See the chapter introduction.

9.

a. $(4 + 2 - 1) \times 2 = 10$	b. $3 \times 5 + 2 \times 4 = 23$	c. $2 \times (4 + 3) + 8 = 22$
d. $50 - (7 - 2) \times 4 = 30$	e. $3 \times 4 - 2 \times 3 = 6$	f. $6 + 7 \times (4 - 2) = 20$

Puzzle Corner:

$16 \times 1 - 1 = 15$ $\quad$ $10 + 5 \times 2 = 20$ $\quad$ $3 + 4 \times 5 + 6 = 29$

$35 - 5 \times 4 = 15$ $\quad$ $5 \times 7 + 6 = 41$ $\quad$ $9 \times 3 - 5 \times 2 = 17$

Mixed Revision Chapter 3, pp. 113-114

Page 113

1. $9 + D = 64$ or $64 - D = 9$; Davy weighs 55 kg.

2.

a. $93 + 6 = 99$ $893 + 6 = 899$	b. $47 + 29 = 76$ $607 + 9 = 616$	c. $15 + 18 = 33$ $624 + 8 = 632$

3. a. 155 $\quad$ b. 327 $\quad$ c. 765

Mixed Revision Chapter 3, cont.

4. a. 801   b. 938   c. 948   d. 657

5.

a. $19 - (6 + 2) + 5 = 16$ $\quad 19 - 6 + 2 + 5 = 20$	b. $(800 - 60) - (50 - 40) = 730$ $\quad 800 - 60 - 50 - 40 = 650$

6. a. 259   b. 176

7. Estimations will vary. For example: 250 m + 250 m + 130 m = 630 m.
   The exact answer is 615 m. This is reasonable because it is close to the estimate.

8. The two jars have 574 beans. If we estimate using rounded numbers, we get $310 + 260 = 570$.
   (As usual, estimations can vary.) The answer of 574 beans is close to the estimate, so it is reasonable.

9.

a. a toy, $28, and a set of books, $129
    toy about $30
    set of books about $130
    together about $160
	b. a ladder, $62, and wheelbarrow, $137
	    ladder about $60
	    wheelbarrow about $140
	    together about $200

Revision Chapter 3, pp. 115-116

1. a.   b.

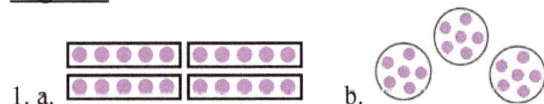

2. a. $7 + 7 + 7 = 21$ b. $20 + 20 + 20 + 20 = 80$

3. a. $5 \times 4 = 20$ b. $9 \times 3 = 27$

4.

a. $2 \times 2 = 4$ $\quad 1 \times 4 = 4$	b. $2 \times 10 = 20$ $\quad 3 \times 3 = 9$	c. $12 \times 0 = 0$ $\quad 12 \times 1 = 12$
d. $0 \times 5 = 0$ $\quad 2 \times 7 = 14$	e. $2 \times 40 = 80$ $\quad 3 \times 30 = 90$	f. $2 \times 400 = 800$ $\quad 1 \times 500 = 500$

5. a. $5 \times 4 = 20$ legs.
   b. $7 - 2 = 5 \times 2 = 10$ cans of cat food.   c. $7 - 1 = 6 \times 3 + 1 = 19$ books total.

6. a. 10   b. 17
   c. 23   d. 12

7.

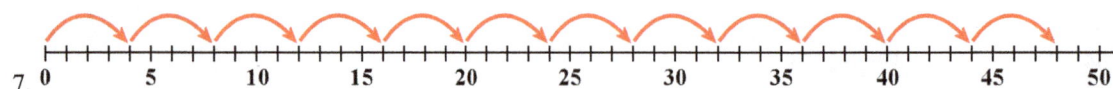

$1 \times 4 = 4$	$4 \times 4 = 16$	$7 \times 4 = 28$	$10 \times 4 = 40$
$2 \times 4 = 8$	$5 \times 4 = 20$	$8 \times 4 = 32$	$11 \times 4 = 44$
$3 \times 4 = 12$	$6 \times 4 = 24$	$9 \times 4 = 36$	$12 \times 4 = 48$

Chapter 4: Multiplication Tables

Multiplication Table of 2, pp. 126-128

1. 0, 2, 4, 6, 8, 10, 12, 14, 16, 18, 20, 22, 24

2. a.

$1 \times 2 = \underline{2}$	$7 \times 2 = \underline{14}$
$2 \times 2 = \underline{4}$	$8 \times 2 = \underline{16}$
$3 \times 2 = \underline{6}$	$9 \times 2 = \underline{18}$
$4 \times 2 = \underline{8}$	$10 \times 2 = \underline{20}$
$5 \times 2 = \underline{10}$	$11 \times 2 = \underline{22}$
$6 \times 2 = \underline{12}$	$12 \times 2 = \underline{24}$

b.

$\underline{1} \times 2 = 2$	$\underline{7} \times 2 = 14$
$\underline{2} \times 2 = 4$	$\underline{8} \times 2 = 16$
$\underline{3} \times 2 = 6$	$\underline{9} \times 2 = 18$
$\underline{4} \times 2 = 8$	$\underline{10} \times 2 = 20$
$\underline{5} \times 2 = 10$	$\underline{11} \times 2 = 22$
$\underline{6} \times 2 = 12$	$\underline{12} \times 2 = 24$

3.

$6 \times 2 = 12$	$7 \times 2 = 14$	$2 \times 3 = 6$	$2 \times 7 = 14$	$2 \times 8 = 16$
$9 \times 2 = 18$	$2 \times 2 = 4$	$2 \times 11 = 22$	$2 \times 4 = 8$	$3 \times 2 = 6$
$4 \times 2 = 8$	$8 \times 2 = 16$	$2 \times 9 = 18$	$2 \times 6 = 12$	$2 \times 5 = 10$
$2 \times 1 = 2$	$12 \times 2 = 24$	$2 \times 12 = 24$	$8 \times 2 = 16$	$10 \times 2 = 20$

4. $\underline{7} \times 2 = 14$ $\underline{6} \times 2 = 12$ $\underline{3} \times 2 = 6$ $\underline{6} \times 2 = 12$ $\underline{11} \times 2 = 22$
$\underline{9} \times 2 = 18$ $\underline{8} \times 2 = 16$ $\underline{9} \times 2 = 18$ $\underline{4} \times 2 = 8$ $\underline{5} \times 2 = 10$
$\underline{4} \times 2 = 8$ $\underline{12} \times 2 = 24$ $\underline{7} \times 2 = 14$ $\underline{10} \times 2 = 20$ $\underline{12} \times 2 = 24$
$\underline{8} \times 2 = 16$ $\underline{1} \times 2 = 2$ $\underline{11} \times 2 = 22$ $\underline{2} \times 2 = 4$ $\underline{3} \times 2 = 6$

5. a. 24, 7, 8   b. 16, 10, 12   c. 18, 0, 2   d. 22, 20, 0

6.

a. Double 8	b. Double 13	c. Double 15
$\underline{8} + \underline{8} = 16$ $\underline{2} \times \underline{8} = 16$	$13 + 13 = 26$ $2 \times 13 = 26$	$15 + 15 = 30$ $2 \times 15 = 30$
d. Double 25	e. Double 32	f. Double 45
$25 + 25 = 50$ $2 \times 25 = 50$	$32 + 32 = 64$ $2 \times 32 = 64$	$45 + 45 = 90$ $2 \times 45 = 90$

7.

$2 \times 12 = 24$ $2 \times 13 = 26$ $2 \times 14 = 28$	$2 \times 15 = 30$ $2 \times 16 = 32$ $2 \times 17 = 34$	$2 \times 18 = 36$ $2 \times 19 = 38$ $2 \times 20 = 40$	$2 \times 21 = 42$ $2 \times 22 = 44$ $2 \times 23 = 46$

8.

a. 14 is even 2×7	b. 7 is odd $2 \times \underline{}$	c. 18 is even 2×9
d. 21 is odd $2 \times$	e. 30 is even 2×15	f. 34 is even 2×17

9. a. $7 \times 2 = 14$
   b. $3 \times 4 = 12$
   c. $5 \times 2 + 4 = 14$
   d. $2 \times 4 + 2 = 10$
   e. $3 \times 4 + 5 \times 2 = 22$

10. Answers will vary. Please check the student's work.

Multiplication Table of 2, cont.

Page 128

11. a. $2 \times 7 - 3 = 11$ Eleven birds stayed in the trees.
    b. $6 \times \$2 - \$8 = \$4$ He has $4 left.

Puzzle Corner: $8 \times \$2 + \$11 = \$27$ The aeroplane cost $27.

Multiplication Table of 4, pp. 129-132

Page 129

1. a. 0, 4, 8, 12, 16, 20, 24, 28, 32, 36, 40, 44, 48
     0, 2, 4, 6, 8, 10, 12, 14, 16, 18, 20, 22, 24

   b. Answers will vary. For example: the skip-counting list by four has every other number from the skip-counting list by 2. Or, the numbers in the top list are doubles of the numbers in the bottom list.

   What do you wonder? Answers will vary. Students might wonder if the same pattern continues on. Or, if you made a list by 6 or 8, if there will be a similar relationship.

2. What do you notice? Answers will vary. For example, that the answers in the table of 4 are doubles of the answers in the table of 2.

$1 \times 2 = 2$	$7 \times 2 = 14$	$1 \times 4 = 4$	$7 \times 4 = 28$
$2 \times 2 = 4$	$8 \times 2 = 16$	$2 \times 4 = 8$	$8 \times 4 = 32$
$3 \times 2 = 6$	$9 \times 2 = 18$	$3 \times 4 = 12$	$9 \times 4 = 36$
$4 \times 2 = 8$	$10 \times 2 = 20$	$4 \times 4 = 16$	$10 \times 4 = 40$
$5 \times 2 = 10$	$11 \times 2 = 22$	$5 \times 4 = 20$	$11 \times 4 = 44$
$6 \times 2 = 12$	$12 \times 2 = 24$	$6 \times 4 = 24$	$12 \times 4 = 48$

3. The shortcut is doubling the number twice.
   a. 52, 64   b. 84, 56   c. 124, 88   d. 412, 204

Page 130

4.   0, 4, 8, 12, 16, 20, 24, 28, 32, 36, 40, 44, 48

5. a.

		b.		
$1 \times 4 = 4$	$7 \times 4 = 28$		$1 \times 4 = 4$	$7 \times 4 = 28$
$2 \times 4 = 8$	$8 \times 4 = 32$		$2 \times 4 = 8$	$8 \times 4 = 32$
$3 \times 4 = 12$	$9 \times 4 = 36$		$3 \times 4 = 12$	$9 \times 4 = 36$
$4 \times 4 = 16$	$10 \times 4 = 40$		$4 \times 4 = 16$	$10 \times 4 = 40$
$5 \times 4 = 20$	$11 \times 4 = 44$		$5 \times 4 = 20$	$11 \times 4 = 44$
$6 \times 4 = 24$	$12 \times 4 = 48$		$6 \times 4 = 24$	$12 \times 4 = 48$

6.

$6 \times 4 = 24$	$7 \times 4 = 28$	$4 \times 3 = 12$	$4 \times 7 = 28$	$3 \times 4 = 12$	$4 \times 8 = 32$
$9 \times 4 = 36$	$8 \times 4 = 32$	$4 \times 11 = 44$	$4 \times 6 = 24$	$4 \times 5 = 20$	$2 \times 4 = 8$
$4 \times 4 = 16$	$12 \times 4 = 48$	$4 \times 9 = 36$	$4 \times 12 = 48$	$10 \times 4 = 40$	$4 \times 1 = 4$

7.

$11 \times 4 = 44$	$3 \times 4 = 12$	$7 \times 4 = 28$	$12 \times 4 = 48$	$6 \times 4 = 24$
$8 \times 4 = 32$	$9 \times 4 = 36$	$11 \times 4 = 44$	$1 \times 4 = 4$	$4 \times 4 = 16$
$2 \times 4 = 8$	$6 \times 4 = 24$	$5 \times 4 = 20$	$10 \times 4 = 40$	$12 \times 4 = 48$

Page 131

8. a. 12   b. 20   c. 30   d. 16
   e. 24   f. 8   g. 18   h. 40

9. The calculations that the students write will vary. Check the students' work. For example:

   a. $3 \times 2 \times 6 = 36$. There are 36 eggs in total.   b. $3 \times 4 + 7 \times 2 = 26$. They have 26 legs in total.

Multiplication Table of 4, cont

10. a. The pattern would continue if the lists were continued.
Why? Answers will vary; check the student's answer.
For example: Because 4 is double 2, and therefore the numbers in the skip-counting list by 4 will always be doubles of the numbers in the skip-counting list by 2.
    b. The numbers in the table of 4 are also even. Student explanations will vary; check the student's work.
For example: The numbers in the table of 4 are also found in the table of 2, so, they are even. Or, when you double an even number, you get an even number.

11. a. $5 \times 2 \times 4 = 40$. They have 40 legs in total.
    b. $5 \times \$3 = \$15$ You can buy five pairs of expensive socks for $15.
    c. $3 \times \$1 + 2 \times \$3 = \$9$. She spent nine dollars.

Puzzle corner. a. $2 \times (5 + 4) + 5 = 23$ b. $30 - 7 \times (2 - 2) = 30$ c. $5 \times (2 + 1) \times 2 = 30$

Multiplication Table of 10, pp. 133-135

1. 0, 10, 20, 30, 40, 50, 60, 70, 80, 90, 100, 110, 120

2. a.

		b.		
$1 \times 10 = 10$	$7 \times 10 = 70$		$1 \times 10 = 10$	$7 \times 10 = 70$
$2 \times 10 = 20$	$8 \times 10 = 80$		$2 \times 10 = 20$	$8 \times 10 = 80$
$3 \times 10 = 30$	$9 \times 10 = 90$		$3 \times 10 = 30$	$9 \times 10 = 90$
$4 \times 10 = 40$	$10 \times 10 = 100$		$4 \times 10 = 40$	$10 \times 10 = 100$
$5 \times 10 = 50$	$11 \times 10 = 110$		$5 \times 10 = 50$	$11 \times 10 = 110$
$6 \times 10 = 60$	$12 \times 10 = 120$		$6 \times 10 = 60$	$12 \times 10 = 120$

...both in the table of two and the table of ten? $10 \times 2 = 2 \times 10$
...both in the table of four and the table of ten? $10 \times 4 = 4 \times 10$

3.

$5 \times 10 = 50$	$6 \times 10 = 60$	$10 \times 8 = 80$	$10 \times 7 = 70$	$2 \times 5 = 10$
$12 \times 10 = 120$	$9 \times 10 = 90$	$10 \times 4 = 40$	$10 \times 10 = 100$	$10 \times 3 = 30$
$7 \times 10 = 70$	$11 \times 10 = 110$	$10 \times 12 = 120$	$10 \times 11 = 110$	$10 \times 6 = 60$

4.

$3 \times 10 = 30$	$2 \times 10 = 20$	$8 \times 10 = 80$	$4 \times 10 = 40$	$9 \times 10 = 90$
$1 \times 10 = 10$	$4 \times 10 = 40$	$9 \times 10 = 90$	$11 \times 10 = 110$	$3 \times 10 = 30$
$6 \times 10 = 60$	$5 \times 10 = 50$	$10 \times 10 = 100$	$7 \times 10 = 70$	$12 \times 10 = 120$

5.

$10 \times 10 = 100$	$15 \times 10 = 150$	$20 \times 10 = 200$
$11 \times 10 = 110$	$16 \times 10 = 160$	$21 \times 10 = 210$
$12 \times 10 = 120$	$17 \times 10 = 170$	$22 \times 10 = 220$
$13 \times 10 = 130$	$18 \times 10 = 180$	$23 \times 10 = 230$
$14 \times 10 = 140$	$19 \times 10 = 190$	$24 \times 10 = 240$

What pattern do you notice? Answers will vary. Check the student's answer. For example: We can just add or tag a zero to the end of the number that is being multiplied by 10.

6. a. 270, 750   b. 480, 500   c. 600, 1000

7. a. 80   b. 70   c. 60   d. 16
   e. 90   f. 50   g. 60   h. 40
   i. 16   j. 120   k. 100   l. 110

Page 135

8. a. 48   b. 16   c. 7   d. 0   e. 20   f. 44

9. It could be 2 cats and 7 chickens, or 3 cats and 5 chickens, or 4 cats and 3 chickens.
   (You cannot really have just 1 cat or 1 chicken since it speaks of them in the plural.)

10.

×	0	1	2	3	4	5	6	7	8	9	10	11	12
0	0	0	0	0	0	0	0	0	0	0	0	0	0
1	0	1	2	3	4	5	6	7	8	9	10	11	12
2	0	2	4	6	8	10	12	14	16	18	20	22	24
3	0	3	6		12						30		
4	0	4	8	12	16	20	24	28	32	36	40	44	48
5	0	5	10		20						50		
6	0	6	12		24						60		
7	0	7	14		28						70		
8	0	8	16		32						80		
9	0	9	18		36						90		
10	0	10	20	30	40	50	60	70	80	90	100	110	120
11	0	11	22		44						110		
12	0	12	24		48						120		

Multiplication Table of 5, pp. 136-138

Page 136

1. 0, 5, 10, 15, 20, 25, 30, 35, 40, 45, 50, 55, 60

2. a.

$1 \times 5 = 5$	$7 \times 5 = 35$
$2 \times 5 = 10$	$8 \times 5 = 40$
$3 \times 5 = 15$	$9 \times 5 = 45$
$4 \times 5 = 20$	$10 \times 5 = 50$
$5 \times 5 = 25$	$11 \times 5 = 55$
$6 \times 5 = 30$	$12 \times 5 = 60$

b.

$1 \times 5 = 5$	$7 \times 5 = 35$
$2 \times 5 = 10$	$8 \times 5 = 40$
$3 \times 5 = 15$	$9 \times 5 = 45$
$4 \times 5 = 20$	$10 \times 5 = 50$
$5 \times 5 = 25$	$11 \times 5 = 55$
$6 \times 5 = 30$	$12 \times 5 = 60$

...both in the table of five and table of two? $2 \times 5 = 5 \times 2$
...both in the table of five and table of four? $4 \times 5 = 5 \times 4$
...both in the table of five and table of ten? $10 \times 5 = 5 \times 10$

3.

$6 \times 5 = 30$	$7 \times 5 = 35$	$5 \times 3 = 15$	$5 \times 7 = 35$	$5 \times 10 = 50$
$9 \times 5 = 45$	$12 \times 5 = 60$	$5 \times 11 = 55$	$5 \times 4 = 20$	$3 \times 5 = 15$
$4 \times 5 = 20$	$8 \times 5 = 40$	$5 \times 9 = 45$	$5 \times 6 = 30$	$5 \times 5 = 25$

4.

$7 \times 5 = 35$	$4 \times 5 = 20$	$11 \times 5 = 55$	$8 \times 5 = 40$	$11 \times 5 = 55$
$1 \times 5 = 5$	$9 \times 5 = 45$	$5 \times 5 = 25$	$10 \times 5 = 50$	$6 \times 5 = 30$
$12 \times 5 = 60$	$2 \times 5 = 10$	$7 \times 5 = 35$	$12 \times 5 = 60$	$3 \times 5 = 15$

Page 137

5. a. 0, 5, 10, 15, 20, 25, 30, 35, 40, 45, 50, 55, 60
   0, 10, 20, 30, 40, 50, 60, 70, 80, 90, 100, 110, 120

   b. Answers will vary. For example: Each number in the list for 10 is a double of the number from the list for 5.

c.

20	30	40	50	60
$\underline{2} \times 10$	$\underline{3} \times 10$	$\underline{4} \times 10$	$\underline{5} \times 10$	$\underline{6} \times 10$
$\underline{4} \times 5$	$\underline{6} \times 5$	$\underline{8} \times 5$	$\underline{10} \times 5$	$\underline{12} \times 5$

   Answers will vary. For example: The bottom numbers are doubles of the top numbers.
   Or, as the first number gets doubled, the second number gets halved.

d.

80	90	120	150	180
$\underline{8} \times 10$	$\underline{9} \times 10$	$\underline{12} \times 10$	$\underline{15} \times 10$	$\underline{18} \times 10$
$\underline{16} \times 5$	$\underline{18} \times 5$	$\underline{24} \times 5$	$\underline{30} \times 5$	$\underline{36} \times 5$

   e. You can multiply by 10 first, and take half of that result. For example 14×5 is half of 14×10, or half of 140.
   So, it is 70. And 28×5 is half of $28 \times 10 = 280$, so it is 140.

6. a. 260, 130   b. 360, 180   c. 840, 420

Page 138

7.

×	0	1	2	3	4	5	6	7	8	9	10	11	12
0	0	0	0	0	0	0	0	0	0	0	0	0	0
1	0	1	2	3	4	5	6	7	8	9	10	11	12
2	0	2	4	6	8	10	12	14	16	18	20	22	24
3	0	3	6		12	15					30		
4	0	4	8	12	16	20	24	28	32	36	40	44	48
5	0	5	10	15	20	25	30	35	40	45	50	55	60
6	0	6	12		24	30					60		
7	0	7	14		28	35					70		
8	0	8	16		32	40					80		
9	0	9	18		36	45					90		
10	0	10	20	30	40	50	60	70	80	90	100	110	120
11	0	11	22		44	55					110		
12	0	12	24		48	60					120		

Puzzle Corner:

5	×	4	= 20
×		×	
2	×	10	= 20
=		=	
10		40	

3	×	4	= 12
×		×	
2	×	6	= 12
=		=	
6		24	

41

More Practice and Revision, pp. 139-140

1. a. 18, 28   b. 10, 12   c. 14, 40   d. 30, 48
   e. 24, 22   f. 24, 30   g. 12, 44   h. 25, 4
   i. 16, 55   j. 60, 36   k. 2, 45   l. 32, 35

2. The equations students write will vary.
   a. $2 \times 12 - 4 = 20$. There are 20 eggs left.
   b. $11 \times 3 + 2 \times 9 = 33 + 18 = 51$. There are 51 workers in total.
   c. $5 \times 4 = 20$. She got five groups.
   d. $4 \times 5 + 3 = 23$. She packed 23 figurines.

5. The operation(s) to be done first is highlighted.

a. $3 + 7 \times 5 = 38$	b. $10 \times 6 - 10 \times 3 = 30$	c. $5 \times (5 - 4) = 5$
d. $(4 + 2) \times 5 = 30$	e. $5 \times 4 + 12 \times 4 = 68$	f. $0 + 7 \times 2 - 4 = 10$

4. a.

9	12	15	18	21	24	27	30	33

b.

20	40	60	80	100	120	140	160	180

c.

48	44	40	36	32	28	24	20	16

5.

×	2	6	9
4	8	24	36
5	10	30	45
10	20	60	90

×	4	2	8	11
3	12	6	24	33
7	28	14	56	77
4	16	8	32	44

Puzzle Corner: a. ☐ = 3, △ = 5 (or vice versa) b. ☐ = 12, △ = 2. c. ☐ = 4, △ = 6 (or vice versa).

Multiplication Table of 3, pp. 141-142

1. 0, 3, 6, 9, 12, 15, 18, 21, 24, 27, 30, 33, 36

2. a.

$1 \times 3 = 3$	$7 \times 3 = 21$
$2 \times 3 = 6$	$8 \times 3 = 24$
$3 \times 3 = 9$	$9 \times 3 = 27$
$4 \times 3 = 12$	$10 \times 3 = 30$
$5 \times 3 = 15$	$11 \times 3 = 33$
$6 \times 3 = 18$	$12 \times 3 = 36$

b.

$1 \times 3 = 3$	$7 \times 3 = 21$
$2 \times 3 = 6$	$8 \times 3 = 24$
$3 \times 3 = 9$	$9 \times 3 = 27$
$4 \times 3 = 12$	$10 \times 3 = 30$
$5 \times 3 = 15$	$11 \times 3 = 33$
$6 \times 3 = 18$	$12 \times 3 = 36$

3.

$6 \times 3 = 18$	$7 \times 3 = 21$	$3 \times 3 = 9$	$3 \times 7 = 21$	$3 \times 8 = 24$
$9 \times 3 = 27$	$2 \times 3 = 6$	$3 \times 11 = 33$	$3 \times 4 = 12$	$3 \times 3 = 9$
$4 \times 3 = 12$	$8 \times 3 = 24$	$3 \times 9 = 27$	$3 \times 6 = 18$	$3 \times 5 = 15$
$3 \times 1 = 3$	$12 \times 3 = 36$	$3 \times 12 = 36$	$8 \times 3 = 24$	$10 \times 3 = 30$

4.

$5 \times 3 = 15$	$4 \times 3 = 12$	$9 \times 3 = 27$	$12 \times 3 = 36$	$10 \times 3 = 30$
$11 \times 3 = 33$	$12 \times 3 = 36$	$11 \times 3 = 33$	$1 \times 3 = 3$	$2 \times 3 = 6$
$3 \times 3 = 9$	$8 \times 3 = 24$	$9 \times 3 = 27$	$6 \times 3 = 18$	$7 \times 3 = 21$

Page 142

5. a. $11 \times 3 + 1 = 34$. Mum is 34 years old.
   b. $10 \times 3 + 1 = 31$. He would have to buy 10 bunches of three and one extra rose.
   c. Answers will vary.

6. a. 24   b. 27   c. 80

7.

×	1	2	3	4	5	6	7	8	9	10	11	12
1	1	2	3	4	5	6	7	8	9	10	11	12
2	2	4	6	8	10	12	14	16	18	20	22	24
3	3	6	9	12	15	18	21	24	27	30	33	36
4	4	8	12	16	20	24	28	32	36	40	44	48
5	5	10	15	20	25	30	35	40	45	50	55	60
6	6	12	18	24	30					60		
7	7	14	21	28	35					70		
8	8	16	24	32	40					80		
9	9	18	27	36	45					90		
10	10	20	30	40	50	60	70	80	90	100	110	120
11	11	22	33	44	55					110		
12	12	24	36	48	60					120		

Multiplication Table of 6, pp. 143-145

Page 143

1. 0, 6, 12, 18, 24, 30, 36, 42, 48, 54, 60, 66, 72

2. a.

$1 \times 6 = 6$	$7 \times 6 = 42$
$2 \times 6 = 12$	$8 \times 6 = 48$
$3 \times 6 = 18$	$9 \times 6 = 54$
$4 \times 6 = 24$	$10 \times 6 = 60$
$5 \times 6 = 30$	$11 \times 6 = 66$
$6 \times 6 = 36$	$12 \times 6 = 72$

b.

$1 \times 6 = 6$	$7 \times 6 = 42$
$2 \times 6 = 12$	$8 \times 6 = 48$
$3 \times 6 = 18$	$9 \times 6 = 54$
$4 \times 6 = 24$	$10 \times 6 = 60$
$5 \times 6 = 30$	$11 \times 6 = 66$
$6 \times 6 = 36$	$12 \times 6 = 72$

3.

$9 \times 6 = 54$	$8 \times 6 = 48$	$6 \times 8 = 48$	$6 \times 5 = 30$	$3 \times 6 = 18$
$2 \times 6 = 12$	$10 \times 6 = 60$	$6 \times 12 = 72$	$6 \times 7 = 42$	$6 \times 6 = 36$
$4 \times 6 = 24$	$3 \times 6 = 18$	$6 \times 9 = 54$	$6 \times 2 = 12$	$6 \times 4 = 24$
$11 \times 6 = 66$	$12 \times 6 = 72$	$6 \times 11 = 66$	$6 \times 6 = 36$	$7 \times 6 = 42$

4.

$12 \times 6 = 72$	$3 \times 6 = 18$	$9 \times 6 = 54$	$7 \times 6 = 42$	$9 \times 6 = 54$
$1 \times 6 = 6$	$8 \times 6 = 48$	$4 \times 6 = 24$	$6 \times 6 = 36$	$5 \times 6 = 30$
$10 \times 6 = 60$	$2 \times 6 = 12$	$7 \times 6 = 42$	$11 \times 6 = 66$	$12 \times 6 = 72$

Page 144

5. a.

Number		Times 3		Times 6
6	=	2 × 3	=	1 × 6
12	=	4 × 3	=	2 × 6
18	=	6 × 3	=	3 × 6
24	=	8 × 3	=	4 × 6

Number		Times 3		Times 6
30	=	10 × 3	=	5 × 6
36	=	12 × 3	=	6 × 6
42	=	14 × 3	=	7 × 6
48	=	16 × 3	=	8 × 6

Answers will vary; check the student's answer. For example: a number times 3 is equal to half that number times 6. Or, the missing numbers in the purple boxes are double the corresponding numbers in the pink boxes.

b. You could double the answer: $42 \times 2 = 84$

6. See the chapter introduction.

7. 4, 9, 9   b. 7, 12, 28   c. 42, 12, 8   d. 6, 45, 36

8. a. $5 \times \$5 + \$10 = \$35$. He has $35 now.
   b. $\$12 + 4 \times \$5 = \$32$. He has enough money to buy the book for $32.

Page 145

9. a.

$$5 \times 1 + 1 = 6$$
$$5 \times 2 + 2 = 12$$
$$5 \times 3 + 3 = 18$$
$$5 \times 4 + 4 = 24$$
$$5 \times 5 + 5 = 30$$
$$5 \times 6 + 6 = 36$$
$$5 \times 7 + 7 = 42$$
$$5 \times 8 + 8 = 48$$
$$5 \times 9 + 9 = 54$$
$$5 \times 10 + 10 = 60$$

b. The list forms a skip-counting pattern by 6, because when you multiply a number by 5, and then add that number, this means you're taking that number 6 times.

10.

×	2	3	4	5	6	7	8	9	10	11	12
2	4	6	8	10	12	14	16	18	20	22	24
3	6	9	12	15	18	21	24	27	30	33	36
4	8	12	16	20	24	28	32	36	40	44	48
5	10	15	20	25	30	35	40	45	50	55	60
6	12	18	24	30	36	42	48	54	60	66	72
7	14	21	28	35	42				70		
8	16	24	32	40	48				80		
9	18	27	36	45	54				90		
10	20	30	40	50	60	70	80	90	100	110	120
11	22	33	44	55	66				110		
12	24	36	48	60	72				120		

1. 0, 11, 22, 33, 44, 55, 66, 77, 88, 99, 110, 121, 132

2. a.

$1 \times 11 = 11$	$7 \times 11 = 77$
$2 \times 11 = 22$	$8 \times 11 = 88$
$3 \times 11 = 33$	$9 \times 11 = 99$
$4 \times 11 = 44$	$10 \times 11 = 110$
$5 \times 11 = 55$	$11 \times 11 = 121$
$6 \times 11 = 66$	$12 \times 11 = 132$

b.

$1 \times 11 = 11$	$7 \times 11 = 77$
$2 \times 11 = 22$	$8 \times 11 = 88$
$3 \times 11 = 33$	$9 \times 11 = 99$
$4 \times 11 = 44$	$10 \times 11 = 110$
$5 \times 11 = 55$	$11 \times 11 = 121$
$6 \times 11 = 66$	$12 \times 11 = 132$

3.

$5 \times 11 = 55$	$2 \times 11 = 22$	$11 \times 7 = 77$	$11 \times 3 = 33$	$11 \times 5 = 55$
$12 \times 11 = 132$	$8 \times 11 = 88$	$11 \times 12 = 132$	$11 \times 10 = 110$	$11 \times 11 = 121$
$9 \times 11 = 99$	$7 \times 11 = 77$	$11 \times 4 = 44$	$11 \times 4 = 44$	$11 \times 9 = 99$
$3 \times 11 = 33$	$6 \times 11 = 66$	$11 \times 11 = 121$	$11 \times 8 = 88$	$11 \times 6 = 66$

4.

$8 \times 11 = 88$	$7 \times 11 = 77$	$5 \times 11 = 55$	$6 \times 11 = 66$	$1 \times 11 = 11$
$12 \times 11 = 132$	$11 \times 11 = 121$	$3 \times 11 = 33$	$2 \times 11 = 22$	$4 \times 11 = 44$
$10 \times 11 = 110$	$9 \times 11 = 99$	$12 \times 11 = 132$	$11 \times 11 = 121$	$10 \times 11 = 110$

5. a.

$8 \times 5 = 40$
$10 \times 5 = 50$
$12 \times 5 = 60$
$14 \times 5 = 70$
$16 \times 5 = 80$
$18 \times 5 = 90$

$20 \times 5 = 100$
$22 \times 5 = 110$
$24 \times 5 = 120$
$26 \times 5 = 130$
$28 \times 5 = 140$
$30 \times 5 = 150$

b. The answers form the skip-counting pattern by 10s. It works that way because we are taking every other multiple of 5. In other words, we multiply 5 by every other number (by 8, by 10, by 12, and so on).

6. a.

×	6	3	4	5
11	66	33	44	55
7	42	21	28	35
9	54	27	36	45
4	24	12	16	20

b.

×	4	5	6	9
2	8	10	12	18
5	20	25	30	45
10	40	50	60	90
4	16	20	24	36

7. Answers will vary. Possible answers are listed, and each of these facts can be written with the factors in the other order.

a.	b.	c.
$1 \times 20 = 20$	$1 \times 18 = 18$	$1 \times 36 = 36$
$4 \times 5 = 20$	$3 \times 6 = 18$	$2 \times 18 = 36$
$2 \times 10 = 20$	$2 \times 9 = 18$	$3 \times 12 = 36$
		$4 \times 9 = 36$
		$6 \times 6 = 36$

8.

×	1	2	3	4	5	6	7	8	9	10	11	12
1	1	2	3	4	5	6	7	8	9	10	11	12
2	2	4	6	8	10	12	14	16	18	20	22	24
3	3	6	9	12	15	18	21	24	27	30	33	36
4	4	8	12	16	20	24	28	32	36	40	44	48
5	5	10	15	20	25	30	35	40	45	50	55	60
6	6	12	18	24	30	36	42	48	54	60	66	72
7	7	14	21	28	35	42				70	77	
8	8	16	24	32	40	48				80	88	
9	9	18	27	36	45	54				90	99	
10	10	20	30	40	50	60	70	80	90	100	110	120
11	11	22	33	44	55	66	77	88	99	110	121	132
12	12	24	36	48	60	72				120	132	

Mystery Number:
a. 28   b. 11 or 121   c. 25

Partial Products, pp. 149-151

1. a.   _4_ groups of 3 plus _2_ groups of 3
       = _6_ groups of 3
   b. $4 \times 3 + 2 \times 3 = 6 \times 3$

2.

a. $4 \times 5 + 3 \times 5$
    $= 7 \times 5$
	b. $1 \times 4 + 4 \times 4$
	    $= 5 \times 4$

c. What single multiplication is equal to that? _5 × 7_	

Partial Products, cont.

3.

a. $5 \times 3 + 5 \times 5$ $= 5 \times 8$	b. $1 \times 8 + 4 \times 8$ $= 5 \times 8$
c. $6 \times 4 + 6 \times 7$ $= 6 \times 11$	d. $2 \times 11 + 4 \times 11$ $= 6 \times 11$

4. 12×10

5. They are both correct. The array shows 7 columns of dots, 6 in each column (7×6 or 6×7). This can be broken down as $2 \times 7 + 4 \times 7$ or as $6 \times 3 + 6 \times 4$.

6. Answers will vary; these are just examples.

a. 14×5 $10 \times 5 + 4 \times 5$ $50 + 20 = \underline{70}$	b. 13×8 $4 \times 8 + 9 \times 8$ $32 + 72 = \underline{104}$
c. 15×6 $10 \times 6 + 5 \times 6$ $60 + 30 = \underline{90}$	d. 16×4 $8 \times 4 + 8 \times 4$ $32 + 32 = \underline{64}$
e. 13×7 $5 \times 7 + 8 \times 7$ $35 + 56 = \underline{91}$	f. 19×5 $10 \times 5 + 9 \times 5$ $50 + 45 = \underline{95}$

Multiplication Table of 9, pp. 152-156

1.

a. 7×9 $7 \times 10 - 7 \times 1$ $70 - 7$ 63	b. 12×9 $12 \times 10 - 12 \times 1$ $120 - 12$ 108	c. 9×9 $10 \times 9 - 1 \times 9$ $90 - 9$ 81
d. 9×6 $60 - 6 = \underline{54}$	e. 4×9 $40 - 4 = \underline{36}$	f. 9×5 $50 - 5 = \underline{45}$

2.

a. 14×9 $140 - 14$ 126	b. 9×17 $170 - 17$ 153	c. 9×23 $230 - 23$ 207
d. 9×35 $350 - 35$ 315	e. 16×9 $160 - 16$ 144	f. 9×56 $560 - 56$ 504

Page 153

3. The orange numbers count up from 0 to 9.
   The blue numbers count down from 9 to 0.

4. 0, 9, 18, 27, 36, 45, 54, 63, 72, 81, 90, 99, 108

$1 \times 9 = 0\ 9$
$2 \times 9 = 1\ 8$
$3 \times 9 = 2\ 7$
$4 \times 9 = 3\ 6$
$5 \times 9 = 4\ 5$
$6 \times 9 = 5\ 4$
$7 \times 9 = 6\ 3$
$8 \times 9 = 7\ 2$
$9 \times 9 = 8\ 1$
$10 \times 9 = 9\ 0$

5. a.

$1 \times 9 = 9$	$7 \times 9 = 63$
$2 \times 9 = 18$	$8 \times 9 = 72$
$3 \times 9 = 27$	$9 \times 9 = 81$
$4 \times 9 = 36$	$10 \times 9 = 90$
$5 \times 9 = 45$	$11 \times 9 = 99$
$6 \times 9 = 54$	$12 \times 9 = 108$

b.

$1 \times 9 = 9$	$7 \times 9 = 63$
$2 \times 9 = 18$	$8 \times 9 = 72$
$3 \times 9 = 27$	$9 \times 9 = 81$
$4 \times 9 = 36$	$10 \times 9 = 90$
$5 \times 9 = 45$	$11 \times 9 = 99$
$6 \times 9 = 54$	$12 \times 9 = 108$

6.

$5 \times 9 = 45$	$8 \times 9 = 72$	$9 \times 10 = 90$	$9 \times 5 = 45$	$9 \times 8 = 72$	$11 \times 9 = 99$
$9 \times 9 = 81$	$10 \times 9 = 90$	$9 \times 3 = 27$	$9 \times 7 = 63$	$1 \times 9 = 9$	$9 \times 2 = 18$
$12 \times 9 = 108$	$6 \times 9 = 54$	$9 \times 1 = 9$	$9 \times 4 = 36$	$9 \times 6 = 54$	$9 \times 9 = 81$

Page 154

7.

$2 \times 9 = 18$	$4 \times 9 = 36$	$8 \times 9 = 72$	$12 \times 9 = 108$	$9 \times 9 = 81$
$5 \times 9 = 45$	$1 \times 9 = 9$	$10 \times 9 = 90$	$11 \times 9 = 99$	$8 \times 9 = 72$
$3 \times 9 = 27$	$8 \times 9 = 72$	$9 \times 9 = 81$	$7 \times 9 = 63$	$6 \times 9 = 54$

8.

Multiply:	Add the digits:
$1 \times 9 = 9$	$0 + 9 = 9$
$2 \times 9 = 18$	$1 + 8 = 9$
$3 \times 9 = 27$	$2 + 7 = 9$
$4 \times 9 = 36$	$3 + 6 = 9$
$5 \times 9 = 45$	$4 + 5 = 9$
$6 \times 9 = 54$	$5 + 4 = 9$
$7 \times 9 = 63$	$6 + 3 = 9$
$8 \times 9 = 72$	$7 + 2 = 9$

Multiply:	Add the digits:
$9 \times 9 = 81$	$8 + 1 = 9$
$10 \times 9 = 90$	$9 + 0 = 9$
$11 \times 9 = 99$	$9 + 9 = 18$; $1 + 8 = 9$
$12 \times 9 = 108$	$1 + 0 + 8 = 9$
$13 \times 9 = 117$	$1 + 1 + 7 = 9$
$14 \times 9 = 126$	$1 + 2 + 6 = 9$
$15 \times 9 = 135$	$1 + 3 + 5 = 9$
$16 \times 9 = 144$	$1 + 4 + 4 = 9$

9. Answers will vary, based on the number chosen. The answer should end up the same as the original number.

Page 155

10. a. **Table of 3:** ⓪, 3 , 6 ,⑨, 12, 15,⑱, 21, 24,㉗,

30, 33,㊱, 39, 42,㊺, 48, 51,(54), 57, 60,

(63), 66, 69,(72)

Table of 9: ⓪,⑨,⑱,㉗, ㊱, ㊺, (54), (63),(72)

Every number in the table of 9 is triple the corresponding number in the table of 3.

b.

27	36	45	54	72
3 × 9	4 × 9	5 × 9	6 × 9	8 × 9
9 × 3	12 × 3	15 × 3	18 × 3	24 × 3

c. 33 × 3 = 99
d. If 108 = 12 × 9, then 108 = 36 × 3.

11. See the chapter introduction.

12. You can find out how many flowers Ava has drawn in total. This is 4 × 9 = 36 flowers. Or, you could find out how many flowers Ava had drawn when she had three rows of flowers. That was 3 × 9 = 27 flowers.

Page 156

13.

×	1	2	3	4	5	6	7	8	9	10	11	12
1	1	2	3	4	5	6	7	8	9	10	11	12
2	2	4	6	8	10	12	14	16	18	20	22	24
3	3	6	9	12	15	18	21	24	27	30	33	36
4	4	8	12	16	20	24	28	32	36	40	44	48
5	5	10	15	20	25	30	35	40	45	50	55	60
6	6	12	18	24	30	36	42	48	54	60	66	72
7	7	14	21	28	35	42			63	70	77	
8	8	16	24	32	40	48			72	80	88	
9	9	18	27	36	45	54	63	72	81	90	99	108
10	10	20	30	40	50	60	70	80	90	100	110	120
11	11	22	33	44	55	66	77	88	99	110	121	132
12	12	24	36	48	60	72			108	120	132	

Puzzle Corner: The comparison you cannot do is marked with a question mark. Since we do not know the value of △, then △ × 1, which equals △, could be more, less, or equal to 70.

9 × △ [<] 10 × △	△ × 5 [>] △ × 4	△ × 0 [<] 3 × 6
△ × 8 [>] △ × 4	4 × △ [<] △ × 8	△ × 1 [?] 10 × 7
△ × 8 [>] △ × 5	△ × 2 [=] △ + △	△ × 3 [>] △ + △

Page 157

1. 0, 7, 14, 21, 28, 35, 42, 49, 56, 63, 70, 77, 84

2. a.

		b.		
$1 \times 7 = 7$	$7 \times 7 = 49$		$1 \times 7 = 7$	$7 \times 7 = 49$
$2 \times 7 = 14$	$8 \times 7 = 56$		$2 \times 7 = 14$	$8 \times 7 = 56$
$3 \times 7 = 21$	$9 \times 7 = 63$		$3 \times 7 = 21$	$9 \times 7 = 63$
$4 \times 7 = 28$	$10 \times 7 = 70$		$4 \times 7 = 28$	$10 \times 7 = 70$
$5 \times 7 = 35$	$11 \times 7 = 77$		$5 \times 7 = 35$	$11 \times 7 = 77$
$6 \times 7 = 42$	$12 \times 7 = 84$		$6 \times 7 = 42$	$12 \times 7 = 84$

3.

$9 \times 7 = 63$	$8 \times 7 = 56$	$7 \times 8 = 56$	$7 \times 5 = 35$	$3 \times 7 = 21$
$4 \times 7 = 28$	$10 \times 7 = 70$	$7 \times 12 = 84$	$7 \times 7 = 49$	$6 \times 7 = 42$
$11 \times 7 = 77$	$7 \times 6 = 42$	$7 \times 9 = 63$	$7 \times 2 = 14$	$4 \times 7 = 28$
$5 \times 7 = 35$	$10 \times 7 = 70$	$6 \times 7 = 42$	$4 \times 7 = 28$	$8 \times 7 = 56$
$11 \times 7 = 77$	$3 \times 7 = 21$	$8 \times 7 = 56$	$12 \times 7 = 84$	$7 \times 7 = 49$
$6 \times 7 = 42$	$2 \times 7 = 14$	$5 \times 7 = 35$	$5 \times 7 = 35$	$9 \times 7 = 63$

4. See the chapter introduction.

Page 158

5. Answers will vary because it is possible to break each multiplication into parts in several different ways. These are just examples. The final answer, naturally, does not vary.

a. 13×7	b. 17×7
$10 \times 7 + 3 \times 7$	$10 \times 7 + 7 \times 7$
$70 + 21 = \underline{91}$	$70 + 49 = \underline{119}$
c. 7×15	d. 7×19
$3 \times 15 + 4 \times 15$	$7 \times 10 + 7 \times 9$
$45 + 60 = \underline{105}$	$70 + 63 = \underline{133}$

6. a. 42   b. 40   c. 32

7. a. 56   b. 98   c. 63   d. 105   e. 84

8.

a.	b.	c.
$1 \times 30 = 30$	$1 \times 12 = 12$	$1 \times 24 = 24$
$2 \times 15 = 30$	$2 \times 6 = 12$	$3 \times 8 = 24$
$5 \times 6 = 30$	$3 \times 4 = 12$	$2 \times 12 = 24$
$3 \times 10 = 30$		

Page 159

9. Student equations will vary.

   a. $b \times 7 = 28$; $b = 4$. Jenny needed four boxes.
   b. $E = 3 \times 12 - 8$; $E = 28$. There are 28 eggs left.
   c. $T \times 6 = 30$; $T = 5$. You need five tables.

Multiplication Table of 7, cont.

Page 159

10.

×	1	2	3	4	5	6	7	8	9	10	11	12
1	1	2	3	4	5	6	7	8	9	10	11	12
2	2	4	6	8	10	12	14	16	18	20	22	24
3	3	6	9	12	15	18	21	24	27	30	33	36
4	4	8	12	16	20	24	28	32	36	40	44	48
5	5	10	15	20	25	30	35	40	45	50	55	60
6	6	12	18	24	30	36	42	48	54	60	66	72
7	7	14	21	28	35	42	49	56	63	70	77	84
8	8	16	24	32	40	48	56		72	80	88	
9	9	18	27	36	45	54	63	72	81	90	99	108
10	10	20	30	40	50	60	70	80	90	100	110	120
11	11	22	33	44	55	66	77	88	99	110	121	132
12	12	24	36	48	60	72	84		108	120	132	

Multiplication Table of 8, pp. 160-162

Page 160

1. 0, 8, 16, 24, 32, 40, 48, 56, 64, 72, 80, 88, 96

2. a.

$1 \times 8 = 8$	$7 \times 8 = 56$
$2 \times 8 = 16$	$8 \times 8 = 64$
$3 \times 8 = 24$	$9 \times 8 = 72$
$4 \times 8 = 32$	$10 \times 8 = 80$
$5 \times 8 = 40$	$11 \times 8 = 88$
$6 \times 8 = 48$	$12 \times 8 = 96$

b.

$1 \times 8 = 8$	$7 \times 8 = 56$
$2 \times 8 = 16$	$8 \times 8 = 64$
$3 \times 8 = 24$	$9 \times 8 = 72$
$4 \times 8 = 32$	$10 \times 8 = 80$
$5 \times 8 = 40$	$11 \times 8 = 88$
$6 \times 8 = 48$	$12 \times 8 = 96$

3.
$8 \times 8 = 64$	$9 \times 8 = 72$	$8 \times 4 = 32$	$8 \times 5 = 40$	$8 \times 8 = 64$
$8 \times 6 = 48$	$8 \times 11 = 88$	$8 \times 12 = 96$	$7 \times 8 = 56$	$8 \times 10 = 80$
$3 \times 8 = 24$	$8 \times 6 = 48$	$2 \times 8 = 16$	$8 \times 9 = 72$	$8 \times 6 = 48$

4.
$4 \times 8 = 32$	$3 \times 8 = 24$	$11 \times 8 = 88$	$5 \times 8 = 40$	$8 \times 8 = 64$
$1 \times 8 = 8$	$6 \times 8 = 48$	$9 \times 8 = 72$	$7 \times 8 = 56$	$12 \times 8 = 96$
$8 \times 8 = 64$	$2 \times 8 = 16$	$10 \times 8 = 80$	$6 \times 8 = 48$	$11 \times 8 = 88$

Multiplication Table of 8, cont.

Page 161

5. Table of 4: <u>**0**</u>, 4, <u>**8**</u>, 12, <u>**16**</u>, 20, <u>**24**</u>, 28, <u>**32**</u>, 36, <u>**40**</u>, 44, <u>**48**</u>, 52, <u>**56**</u>, 60, <u>**64**</u>, 68, <u>**72**</u>, 76, <u>**80**</u>, 84, <u>**88**</u>, 92, <u>**96**</u>
   Table of 8: <u>**0**</u>, <u>**8**</u>, <u>**16**</u>, <u>**24**</u>, <u>**32**</u>, <u>**40**</u>, <u>**48**</u>, <u>**56**</u>, <u>**64**</u>, <u>**72**</u>, <u>**80**</u>, <u>**88**</u>, <u>**96**</u>
   What do you notice? Student answers will vary. For example: Every second number in the table of 4 is also in the table of 8. The numbers in the table of 8 are doubles of the numbers in the table of 4.

6.

Number		Times 4		Times 8	Number		Times 4		Times 8
8	=	2 × 4	=	1 × 8	40	=	10 × 4	=	5 × 8
16	=	4 × 4	=	2 × 8	48	=	12 × 4	=	6 × 8
24	=	6 × 4	=	3 × 8	56	=	14 × 4	=	7 × 8
32	=	8 × 4	=	4 × 8	64	=	16 × 4	=	8 × 8

A number times 4 is equal to half that number times 8.
You could double 52 to find that $13 \times 8 = 104$.

7. a. $8 \times 5 = 40$. There are 40 rubbers in five packages.
   b. $8 \times 3 = 24$. She needs three packages of rubbers so each child can have one.
   c. $2 \times 5 = 10$. It will take them five weeks to eat ten kilograms of beans.

Page 162

8.

×	1	2	3	4	5	6	7	8	9	10	11	12
1	1	2	3	4	5	6	7	8	9	10	11	12
2	2	4	6	8	10	12	14	16	18	20	22	24
3	3	6	9	12	15	18	21	24	27	30	33	36
4	4	8	12	16	20	24	28	32	36	40	44	48
5	5	10	15	20	25	30	35	40	45	50	55	60
6	6	12	18	24	30	36	42	48	54	60	66	72
7	7	14	21	28	35	42	49	56	63	70	77	84
8	8	16	24	32	40	48	56	64	72	80	88	96
9	9	18	27	36	45	54	63	72	81	90	99	108
10	10	20	30	40	50	60	70	80	90	100	110	120
11	11	22	33	44	55	66	77	88	99	110	121	132
12	12	24	36	48	60	72	84	96	108	120	132	

Puzzle Corner: a. ▢ = 6, △ = 8 (or vice versa) b. ▢ = 4, △ = 12 (or vice versa). c. ▢ = 12, △ = 3.

Multiplication Table of 12, pp. 163-164

Page 163

1. 0, 12, 24, 36, 48, 60, 72, 84, 96, 108, 120, 132, 144

2. a.

1 × 12 = 12	7 × 12 = 84
2 × 12 = 24	8 × 12 = 96
3 × 12 = 36	9 × 12 = 108
4 × 12 = 48	10 × 12 = 120
5 × 12 = 60	11 × 12 = 132
6 × 12 = 72	12 × 12 = 144

b.

1 × 12 = 12	7 × 12 = 84
2 × 12 = 24	8 × 12 = 96
3 × 12 = 36	9 × 12 = 108
4 × 12 = 48	10 × 12 = 120
5 × 12 = 60	11 × 12 = 132
6 × 12 = 72	12 × 12 = 144

3.

3 × 12 = 36	9 × 12 = 108	12 × 4 = 48	12 × 1 = 12	7 × 12 = 84
2 × 12 = 24	10 × 12 = 120	12 × 5 = 60	12 × 7 = 84	12 × 3 = 36
1 × 12 = 12	6 × 12 = 72	12 × 8 = 96	12 × 9 = 108	4 × 12 = 48
8 × 12 = 96	12 × 12 = 144	12 × 11 = 132	12 × 6 = 72	12 × 2 = 24

4.

3 × 12 = 36	2 × 12 = 24	7 × 12 = 84	6 × 12 = 72	12 × 12 = 144
1 × 12 = 12	4 × 12 = 48	12 × 12 = 144	10 × 12 = 120	11 × 12 = 132
6 × 12 = 72	5 × 12 = 60	8 × 12 = 96	5 × 12 = 60	9 × 12 = 108

Page 164

5. See the chapter introduction.

6.

×	1	2	3	4	5	6	7	8	9	10	11	12
1	1	2	3	4	5	6	7	8	9	10	11	12
2	2	4	6	8	10	12	14	16	18	20	22	24
3	3	6	9	12	15	18	21	24	27	30	33	36
4	4	8	12	16	20	24	28	32	36	40	44	48
5	5	10	15	20	25	30	35	40	45	50	55	60
6	6	12	18	24	30	36	42	48	54	60	66	72
7	7	14	21	28	35	42	49	56	63	70	77	84
8	8	16	24	32	40	48	56	64	72	80	88	96
9	9	18	27	36	45	54	63	72	81	90	99	108
10	10	20	30	40	50	60	70	80	90	100	110	120
11	11	22	33	44	55	66	77	88	99	110	121	132
12	12	24	36	48	60	72	84	96	108	120	132	144

Mixed Revision Chapter 4, pp. 165-166

Page 165

1. a. 660   b. 600   c. 820   d. 60

2. a. 746   b. 721

Mixed Revision Chapter 4, cont.

Page 165

3. The equations the student writes will vary. For example: $16 + x = 52$ or $x = 52 - 16$. Solution: $x = 36$.
   There are 36 white candles.

4. a. The equations the student writes will vary. For example: $P = 5 \times 7 - 1$. Solution: $P = 34$. She got 34 pepper plants.
   b. The equations the student writes will vary. For example: $11 \times 4 + J = 47$ or $J + 11 \times 4 = 47$ or $J = 47 - 4 \times 11$.
   Solution: $J = 3$. Jim has three pencils.

5. a. △ = 27   b. △ = 700   c. △ = 430

Page 166

6.

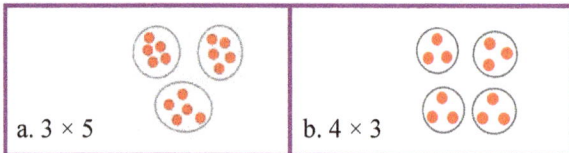

a. 3 × 5   b. 4 × 3

7. a. $5 \times 10 = 50$ Five children have 50 toes in total.
   b. $3 \times 5 = 15$ He got three rows of cars.
   c. $5 \times 4 = 20$ You need five tables to seat 20 people.

8. a. 290, 520   b. 410, 466   c. 433, 739

9. a. 10, 0   b. 12, 0   c. 20, 9   d. 7, 0

10. $5 \times 4 = 20$

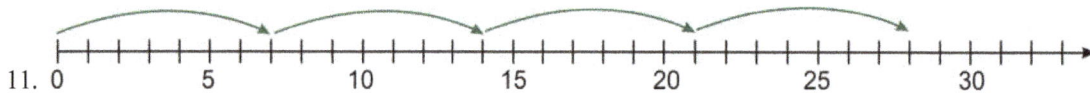

11. 0   5   10   15   20   25   30

Revision Chapter 4, pp. 167-169

Page 167

1.

×	0	1	2	3	4	5	6	7	8	9	10	11	12
0	0	0	0	0	0	0	0	0	0	0	0	0	0
1	0	1	2	3	4	5	6	7	8	9	10	11	12
2	0	2	4	6	8	10	12	14	16	18	20	22	24
3	0	3	6	9	12	15	18	21	24	27	30	33	36
4	0	4	8	12	16	20	24	28	32	36	40	44	48
5	0	5	10	15	20	25	30	35	40	45	50	55	60
6	0	6	12	18	24	30	36	42	48	54	60	66	72
7	0	7	14	21	28	35	42	49	56	63	70	77	84
8	0	8	16	24	32	40	48	56	64	72	80	88	96
9	0	9	18	27	36	45	54	63	72	81	90	99	108
10	0	10	20	30	40	50	60	70	80	90	100	110	120
11	0	11	22	33	44	55	66	77	88	99	110	121	132
12	0	12	24	36	48	60	72	84	96	108	120	132	144

Revision Chapter 4, cont.

Page 167

2. a. 16   b. 90   c. 84

3. The idea that 4 groups of 4, plus 2 groups of four, equals 6 groups of 4. With symbols: $4 \times 4 + 2 \times 4 = 6 \times 4$.

4. 5×7.

Page 168

5.

a. $\underline{7} \times 4 = 28$
   $36 = 4 \times \underline{9}$
   $\underline{7} \times 12 = 84$

b. $108 = 12 \times \underline{9}$
   $32 = \underline{4} \times 8$
   $8 \times \underline{9} = 72$

c. $36 = \underline{12} \times 3$
   $\underline{7} \times 3 = 21$
   $\underline{5} \times 12 = 60$

6. a. <   b. >   c. >
   d. >   e. =   f. >

7. You can double the answer to 17×4 to find 17×8. So, 17×8 is double 68, or 136.

8.

a. $4 \times 5 = 20$ There will be $\underline{5}$ groups.	b. $4 \times \$3 + \$7 = \$19$ The total cost was $\underline{\$19}$.
c. $12 \times \$2 = \$24$ He bought $\underline{12}$ packages.	d. $5 \times 4 + 3 \times 4 + 20 \times 2 = 72$ They have $\underline{72}$ feet in total.

Page 169

9.

×	2	7	6
5	10	35	30
9	18	63	54
11	22	77	66

×	12	3	9	5
11	132	33	99	55
4	48	12	36	20
7	84	21	63	35

10.

a.

96	88	80	72	64	56	48	40	32

b.

60	120	180	240	300	360	420	480	540

c.

18	27	36	45	54	63	72	81	90

Mystery Numbers: a. 44   b. 24   c. 29   d. 24   e. 44   f. 12

Chapter 5: Time

Revision: Reading the Clock, pp. 174-175

Page 174

1.

a.	b.	c.	d.
8 : 15	11 : 30	2 : 40	5 : 55
8 : 20	11 : 35	2 : 45	6 : 00
8 : 25	11 : 40	2 : 50	6 : 05
8 : 30	11 : 45	2 : 55	6 : 10
8 : 35	11 : 50	3 : 00	6 : 15
8 : 40	11 : 55	3 : 05	6 : 20
8 : 45	12 : 00	3 : 10	6 : 25
8 : 50	12 : 05	3 : 15	6 : 30
8 : 55	12 : 10	3 : 20	6 : 35
9 : 00	12 : 15	3 : 25	6 : 40

Page 175

2. a. 1:40 b. 12:05 c. 8:25 d. 3:45
   e. 7:35 f. 11:50 g. 7:20 h. 1:15

3. The time now → | a. 3:35 | b. 5:20 | c. 4:50 | d. 7:45 |

   10 min. later → | 3:45 | 5:30 | 5:00 | 7:55 |

4. 5 min. earlier → a. 3:50   b. 4:00   c. 3:55   d. 6:55

Half and Quarter Hours, pp. 176-178

Page 176

1.

a. a quarter past 8   b. a quarter to 9   c. a quarter past 9   d. a quarter to 8

8:15   8:45   9:15   7:45

2. a. a quarter past 8   b. a quarter to 5   c. 11:45

Half and Quarter Hours, cont.

Page 177

3. b. a quarter past 9
   c. a quarter to 6   d. a quarter to 1

4.

20 past 10   a quarter to 10   20 to 10

half past 10   a quarter past 10   5 to 10

5. a. 8:15   b. 6:50   c. 3:25   d. 5:30
   e. 4:45   f. 5:35   g. 8:45   h. 7:05
   i. 11:45 j. 6:15   k. 10:55 l. 8:40

Page 178

6. a. 5:45 or a quarter to 6
   b. 2:15 or a quarter past 2
   c. 1:45 or a quarter to 2

7. a. half past 7   b. a quarter past 5
   c. a quarter to 6   d. a quarter to 10
   e. a quarter past 12   f. half past 12
   g. a quarter to 12   h. a quarter to 8

Puzzle Corner:

Time now:

Half-hour later: a quarter past 5   4 o'clock   a quarter past 6   a quarter to 8   a quarter to 7

Using to and Past in Telling Time, pp. 179-181

Page 179

1. a. 5 past 4   b. 10 past 11   c. 20 past 3
   d. 25 past 2   e. 10 past 9   f. 25 past 11

2.

a.   20 past 3
     3 : 20

b.   5 past 12
     12 : 05

c.   10 past 7
     7 : 10

57

Using to and Past in Telling Time, cont.

3. a. 5 to 10   b. 20 to 9   c. 25 to 7

4. a. 15 to 5   b. 5 to 2   c. 25 to 4
   d. 20 to 3   e. 10 to 11   f. 25 to 12

5.

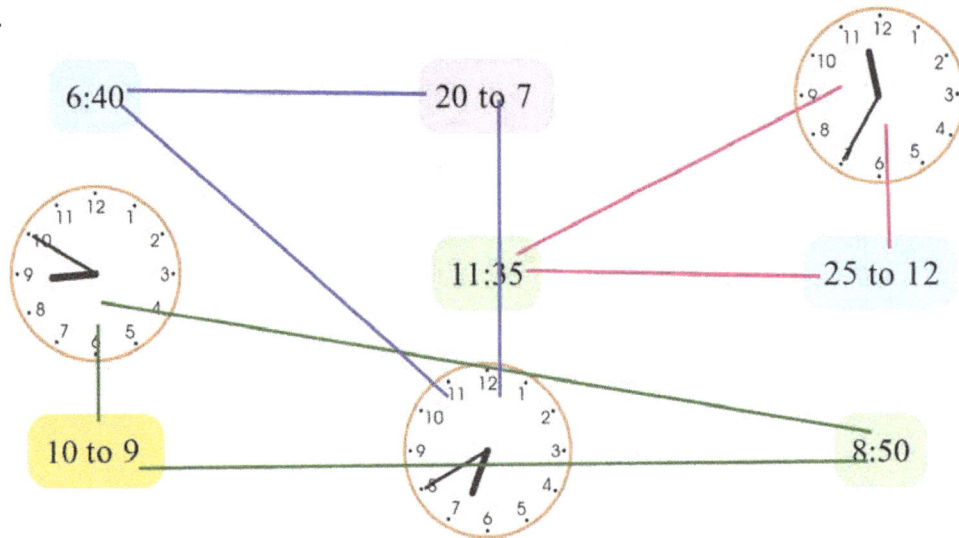

6. 25 minutes

7. 10 minutes

8. a. 7:20   b. 5:10
   c. 10:55   d. 3:40
   e. 8:35   f. 11:50

9. a. 10 to 12   b. 25 to 3   c. 20 past 9
   d. 5 past 8   e. 10 past 12   f. 15 to 4 or quarter to 4
   g. 5 to 9   h. 20 to 1   i. 25 to 2

Clock to the Minute, pp. 182-184

1. a. 1:03   b. 1:24   c. 1:14   d. 1:11

2. See the chapter introduction.

3.

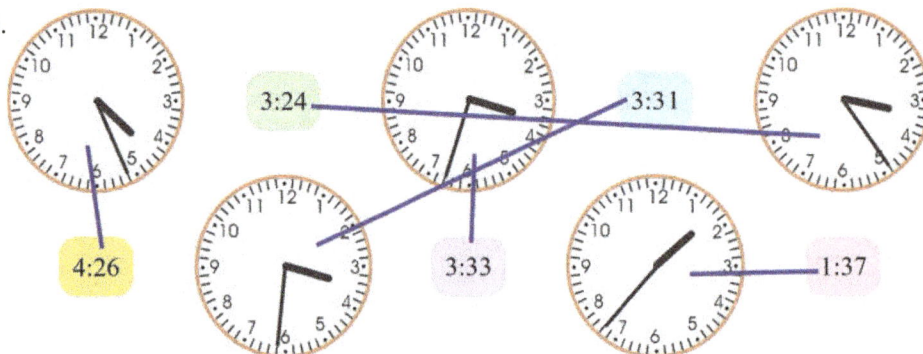

Clock to the Minute, cont.

Page 183

4. a. 1:57   b. 2:47   c. 3:24   d. 4:38
   e. 5:23   f. 8:17   g. 8:43   h. 5:44
   i. 12:41   j. 10:49   k. 9:29   l. 11:11

Page 184

5.

The time now →   a. 10:49   b. 5:24   c. 1:14

10 min. later →   10:59   5:34   1:24

6. a. 8:32   b. 11:26   c. 12:56

7. a. 2:07   b. 12:01   c. 3:08

Elapsed Time 1, pp. 185-186

Page 185

1. See the chapter introduction.

2. a. 15 minutes   b. 15 minutes   c. 20 minutes   d. 35 minutes

3. a. 10 minutes   b. 10 minutes   c. 25 minutes

4. a. 3:30   b. 3:45
   c. 6:40   d. 12:05

Page 186

5. A _8:30_   B _8:50_   C _9:05_   D _9:35_   E _9:50_

6. a. 15 minutes   b. 10 minutes   c. 25 minutes

7. a. 20 minutes   b. 15 minutes   c. 30 minutes

Elapsed Time 2, pp. 187-188

Page 187

1. a. 13 minutes   b. 37 minutes   c. 44 minutes   d. 57 minutes
   e. 27 minutes   f. 36 minutes   g. 14 minutes   h. 23 minutes

Page 188

2. See the chapter introduction.

3. a. 19 minutes   b. 13 minutes   c. 36 minutes   d. 27 minutes

4. a. 22 minutes   b. 49 minutes   c. 35 minutes
   d. 49 minutes   e. 21 minutes   f. 54 minutes

5. a. At 4:52 she should take it out.
   b. It is 1:46 PM.
   c. She should wake up at 5:34 AM.
   d. The class started at 1:45 PM.

Elapsed Time 3, pp. 189-191

Page 189

1. A 11:38   B 11:49   C 12:17   D 12:36

2. 3 + 20 + 10 + 2 = 35 minutes

3. 9 + 10 + 20 + 6 = 45 minutes

Page 190

4. a. 17 minutes   b. 14 minutes   c. 32 minutes

5.

NOW: (clock a.)	NOW: (clock b.)
10 minutes later 6:55	10 minutes earlier 4:15
16 minutes later 7:01	12 minutes earlier 4:13
25 minutes later 7:10	30 minutes earlier 3:55

6. a. 29 minutes   b. 15 minutes   c. 11 minutes

Page 191

7. a. She arrived at the store at 4:10.
   b. She got home at 5:00.
   c. She crocheted for 38 minutes.
   d. He spent 16 minutes on homework.

Puzzle Corner: Starting at 4:10 PM, 25 minutes after that is 4:35. Then, 35 minutes after that is 5:10, and 20 minutes later is 5:30 PM. From 5:30 PM till 6 PM is 30 minutes. So Mateo played with his little brother for <u>30 minutes</u>.

Elapsed Time 4, pp. 192-194

Page 192

1. a. b. c. Answers will vary. Please check the student's work.

2.

a. 40 minutes to 8:00.	c. 40 minutes to 2:30.
b. 23 minutes to 8:00.	d. 33 minutes to 2:30.

Page 193

3.

from	10:30	8:30	1:40	5:45	3:20 AM
to	11:30	12:30	7:40	11:45	12:20 PM
elapsed time	1 hour	4 hours	6 hours	6 hours	9 hours

Elapsed Time 4, cont.

4.

a.	b.
TIME NOW:	TIME NOW:
30 min later 8:15	2 hours earlier 10:49
2 hours later 9:45	1 hour earlier 11:49
5 hours later 12:45	40 min earlier 12:09

5. a. Music class is 32 minutes long.
   b. Sergio started answering emails at 9:35.

6.

from	1:25	1:58	9:05	7:24	5:45
to	1:55	2:15	9:53	8:00	6:14
elapsed time	*30 min*	17 minutes	48 minutes	36 minutes	29 minutes

7.

from	4:20	3:06	7:50	11:10	6:53
to	4:40	3:32	8:03	12:00	7:10
elapsed time	*20 min*	*26 min*	*13 min*	*50 min*	*17 min*

8. When the amount of time that has passed has moved into the next hour, you can't simply subtract the minutes because the smaller number of minutes is the later time. You have to add the amount of minutes until the full hour to the number of minutes past the full hour to find the elapsed time.

9. It was 4:09 PM when Eva got off the bus.

10. a. 45 minutes   b. 39 minutes   c. 20 minutes

Using the Calendar, pp. 195-196

1. See the chapter introduction.

2. a. 2 November   b. 5 December   c. 14 November   d. 6 December
   e. 17 December   f. 22 November   g. 23 January   h. 10 December

3. Camp started on 22 July; the 26th was the last day of camp.

4. 7 December.

5. It is 33 days till Mum's birthday.

6. Mary should return it at the latest on 18 April.

Mixed Revision Chapter 5, pp. 197-198

Page 197

1. a. 25   b. 43   c. 29
   d. 389   e. 561   f. 803

2. a. 27   b. 687   c. 5

3. The cheaper one costs $89 – $17 = $72. The two cheaper ones together cost $72 + $72 = $144.
   If we estimate the cost, the cheaper one costs about $90 – $20 = $70, and two of them about $140.
   The answer of $144 is reasonable because it is close to the estimate.

4. a. 27, 28, 0   b. 24, 24, 24   c. 56, 63, 32   d. 36, 60, 21

5. a. $7 \times 2 = 14$; He read 14 books.
   b. $3 \times 7 + 5 = 26$; He put 26 pencils into the pencil cases.

Page 198

6. a. See the pattern on the right.
   b. The answers go down by 10 each time, because the numbers being
      subtracted go up by 10 each time.

7. The first step is highlighted.

   a. $2 + 5 \times 2 = 12$ b. $5 \times (1 + 1) = 10$ c. $(4 - 2) \times 7 = 14$

$$657 - 10 = \underline{647}$$
$$657 - 20 = \underline{637}$$
$$657 - 30 = \underline{627}$$
$$657 - 40 = \underline{617}$$
$$657 - 50 = \underline{607}$$
$$657 - 60 = \underline{597}$$

8. These equations match the problem: $275 - 48 = x$ or $x + 48 = 275$.
   Solution: $275 - 48 = 227$; The cheaper washer costs $227.

9. 430 + 430 + 280 + 280 = 1420 metres approximately.
   or 400 + 400 + 300 + 300 = 1400 metres approximately.

Revision Chapter 5, pp. 199-200

Page 199

1.

The time now →

| a. 8:43 | b. 4:57 | c. 1:14 |

10 min. later →

| 8:53 | 5:07 | 1:24 |

2. a. 34 minutes   b. 35 minutes

3. a. 35 minutes   b. 38 minutes   c. 11 minutes

Page 200

4.

from	8 AM	7 AM	9 AM	11 AM	10 AM
to	12 noon	1 PM	4 PM	11 PM	7 PM
hours	4	6	7	12	9

5. It ended at 1:57 PM.

6. The trip started at 11:27 AM.

7. Her birthday is 2 April.

8. 5 April 5

9. 14 March

Chapter 6: Money

Counting Coins, pp. 203-204

1. See the chapter introduction.

2. a. 80 cents   b. 220 cents
   c. 560 cents   d. 165 cents

3.

a. four 20-cent coins = 80c	b. six 50-cent coins = $3
five 20-cent coins = 100c = $1	seven 50-cent coins = $3.50
six 20-cent coins = 120c	eight 50-cent coins = $4
seven 20-cent coins = 140c	nine 50-cent coins = $4.50
eight 20-cent coins = 160c	ten 50-cent coins = $5

4.

a. _7_ 20-cent coins = 140 cents	b. _4_ 50-cent coins = $2
_10_ 20-cent coins = 200 cents	_12_ 50-cent coins = $6
_13_ 20-cent coins = 260 cents	_22_ 50-cent coins = $11

5. a. Both are correct.
   b. Answers will vary; check the student's answer. For example: six 20-cent coins and one 10-cent coin or a 1-dollar coin and three 10-cent coins, etc.

6. Answers will vary. The table below shows all the possibilities.

a. 90c	b. 105c
_4_ 20-cent coins + _2_ 5-cent coins	_5_ 20-cent coins + _1_ 5-cent coins
_3_ 20-cent coins + _6_ 5-cent coins	_4_ 20-cent coins + _5_ 5-cent coins
_2_ 20-cent coins + _10_ 5-cent coins	_3_ 20-cent coins + _9_ 5-cent coins
_1_ 20-cent coin + _14_ 5-cent coins	_2_ 20-cent coins + _13_ 5-cent coins
	_1_ 20-cent coin + _17_ 5-cent coins

7. Answers will vary; check the student's work. The student can learn from this exercise that even small coins can eventually add up! Some possibilities, but not all, are shown below.

a. 65c three 20-c coins and one 5-cent coin
   six 10-c coins and one 5-cent coin
   one 50-cent coin and three 5-cent coins

b. 125c one 1-dollar coin and five 5-cent coins
   one 1-dollar coin, a 20-cent coin, and a 5-cent coin
   six 20-cent coins and one 5-cent coin

c. 90c four 20-cent coins and a 10-cent coin
   one 50-cent coin and four 10-cent coins
   one 50-cent coin and two 20-cent coins

d. 260c one 2-dollar coin and three 20-cent coins
   two 1-dollar coins and six 10-cent coins
   five 50-cent coins and one 10-cent coin

Dollars, pp. 205-207

Page 205

1. a. $15.30   b. $92.25   c. $67.20   d. $113.30

Page 206

2. a. $6.10   b. $22.90   c. $13.40

3. a. $0.65   b. $0.40   c. $0.25
   d. $0.80   e. $1.05   f. $0.80

Page 207

4. a. $0.55   b. $0.05   c. $4.25
   d. 565c   e. 30c   f. 305c

5. Mark has $7.45.

6. a. $3.85   b. $1.70   c. $6.35   d. $5.35   e. $1.55   f. $1.45

Counting Up to Make Change, pp. 208-210

Page 208

1. Answers will vary since one can make the change using different coins and notes. Check the student's work.
   a. one 10-cent coin, one dollar; The change is $1.10.
   b. one 10-cent coin, one 5-cent coin, one 50-cent coin; The change is $0.65.

Page 209

2. Answers will vary since one can make the change using different coins and notes. Check the student's work.
   a. one 5-cent coin, four 20-cent coins, a 5-dollar note; The change is $5.85.
   b. One 50-cent coin, one 10-dollar note, one 5-dollar note; The change is $15.50.

3. Answers will vary since one can make the change using different coins and notes. Check the student's work.
   a. One 5-cent coin, two 20-cent coins, one 2-dollar coin. The change is $2.45.
   b. Two 20-cent coins, one 1-dollar coin, two 10-dollar notes. The change is $21.40.
   c. One 50-cent coin, two 2-dollar coins. The change is $4.50.
   d. One 5-cent coin, one 2-dollar coin, one 10-dollar note. The change is $12.05.

Page 210

4. a. Yes, the change is correct.
   b. No, the correct change is $11.80 (10-dollar note, 1-dollar coin, a 50-cent coin, a 20-cent coin, and a 10-cent coin).

5. Answers will vary since one can make the change using different coins and notes. Check the student's work.
   a. Change: $1.70. Use one 1-dollar coin, three 20-cent coins, and one 10-cent coin.
   b. Change: $3.60. Use one 1-dollar coin, one 2-dollar coin, and three 20-cent coins.
   c. Change: $2.80. Use a 2-dollar coin, a 50-cent coin, a 20-cent coin, and a 10-cent coin.
   d. Change: $2.75. Use a 2-dollar coin, a 50-cent coin, a 20-cent coin, and a 5-cent coin.
   e. Change: $13.80. Use one 10-dollar note, three 1-dollar coins, and four 20-cent coins.

Making Change, pp. 211-212

Page 211

1. a. $3.00   b. $6.00   c. $4.50   d. $3.60   e. $2.40   f. $1.35

2. a. No, the correct change is $6.15 (a 5-dollar note, a 1-dollar coin, a 10-cent coin, and a 5-cent coin).
   b. No, the correct change is $2.25 (a 2-dollar coin, a 20-cent coin, and a 5-cent coin).

Page 212

3. a. 44   b. 81   c. 28   d. 56   e. 66

4. a. 65c, $24, $73   b. 55 c, 90c, 15c   c. $81, $52, $51.35

Making Change, cont.

Page 212

5. There are numerous possibilities for the selection of notes and coins used for the change. Please check that what the student has drawn adds up to the correct amount.
   a. Change: $16.45. Use a 10-dollar note, a 5-dollar note, a 1-dollar coin, two 20-cent coins, and a 5-cent coin.
   b. Change: $25.25. Use a 20-dollar note, a 5-dollar note, a 20-cent coin, and a 5-cent coin.
   c. Change: $12.90. Use a 10-dollar note, a 2-dollar coin, four 20-cent coins, and a 10-cent coin.
   d. Change: $17.05. Use a 10-dollar note, a 5-dollar note, a 2-dollar coin, and a 5-cent coin.

Using Mental Maths to Solve Money Problems, pp. 213-214

Page 213

1. a. $6.10   b. $2.00   c. $4.40
   d. $5.30   e. $1.05   f. $8.00
   g. $6.30   h. and i. Answers will vary. Please check the student's totals.

2. a. scissors and rubber   b. pen and stapler   c. crayons and stapler

Page 214

3. See the chapter introduction.

4. a. Change: $3.75   b. Change: $18.65
   c. Change: $0.25   d. Change: $26.50

5. a. Total: $3.90; Change: $0.10
   b. Total: $6.45; Change: $3.55
   c. No, I can't; I need 60 cents more.
   d. Yes, I can, and my change is 50 cents.

Finding the Total and Change, pp. 215-216

Page 215

1. a. $35.45   b. $16.80   c. $35.45   d. $106.95

2. a. $27.40   b. $37.75   c. $77.40

Page 216

3. a. Yes, it was correct. $11.50 + $8.50 = $20.
   b. Yes, it was correct. $35.90 + $14.10 = $50.
   c. Julie can buy 3 pairs of scissors. $5.15 + $5.15 + $5.15 = $15.45, but $5.15 + $5.15 + $5.15 + $5.15 = $20.60.
   d. The total is $15.95 + $5.75 + $5.75 + $5.75 = $33.20.

More Problem Solving, pp. 217-218

Page 217

1.

a. $10 − $2.65	b. $50 − $28.35
+ $0.35 + $7.00	+ $0.65 + $21.00
$2.65 $3.00 $10.00	$28.35 $29.00 $50.00
So, $10 − $2.65 = $7.35	So, $50 − $28.35 = $21.65

2. a. $5.55   b. $12.75   c. $13.55   d. $15.50

More Problem Solving, cont.

Page 217

3. $50 – ($15.55 + $2.40) = $32.05. His change was $32.05.

Page 218

4. a. $35.90 + $8.90 = $44.80 – $25 = $19.80. He needs $19.80 more.
   b. Her change is $2.30. The total cost is $4.55 + $2.30 + $0.85 = $7.70. Change: $10 – $7.70 = $2.30
   c. John's total bill was $9.80, and his change was $10.20.
   d. $14.55 + $23.95 = $38.50. So, yes, she can, and her change is $1.50.

Making Sense with No Cents, pp. 219-221

Page 220

1. a. Rounded price 60c   b. Rounded price 90c   c. Rounded price 65c
   d. Rounded price 95c   e. Rounded price 95c   f. Rounded price 90c

2.

a. $0.67 ≈ $0.65	b. $1.24 ≈ $1.25	c. $8.09 ≈ $8.10	d. $13.04 ≈ $13.05
e. $20.86 ≈ $20.85	f. $4.53 ≈ $4.55	g. $54.28 ≈ $54.30	h. $33.51 ≈ $33.50

3.

a. $1.56 Customer gives $2	rounded price ≈ $1.55	Change: $0.45
b. $6.67 Customer gives $8	rounded price ≈ $6.65	Change: $1.35
c. $1.88 Customer gives $2	rounded price ≈ $1.90	Change: $0.10
d. $4.23 Customer gives $5	rounded price ≈ $4.25	Change: $0.75
e. $0.94 Customer gives $1	rounded price ≈ $0.95	Change: $0.05

Page 221

4.

a. A toy $7.44; Customer gives $10 Rounded price: $7.45; Change $2.55	b. A pencil set $4.61; Customer gives $4.75 Rounded price: $4.60; Change $0.15
c. A book $13.88; Customer gives $14 Rounded price: $13.90; Change $0.10	d. A postcard $0.47; Customer gives $0.50 Rounded price: $0.45; Change $0.05

5. a. $34.91 + $12.32 + $5.88 = $53.11, which rounds down to $53.10.
   b. $0.31 + $0.31 + $0.31 + $1.04 +$1.04 + $4.97 = $7.98, which rounds up to $8.00.
      The change is $10.00 – $8.00 = $2.00.
   c. $18.98 rounds up to $19.00. The change is $20.00 – $19.00 = $1.00.

   d. $44.02 is rounded down to $44.00. The change is $20.00 + $20.00 + $20.00 – $44.00 = $16.00.

Mixed Revision Chapter 6, pp. 222-223

Page 222

1. a. 2:59   b. 4:56   c. 9:09   d. 11:31

2. a. $50 - (20 - 7) = 37$ b. $(8 - 5) \times 2 - 1 = 5$ c. $(15 + 5) \times (2 - 1) = 20$ OR $15 + 5 \times (2 - 1) = 20$

3. a.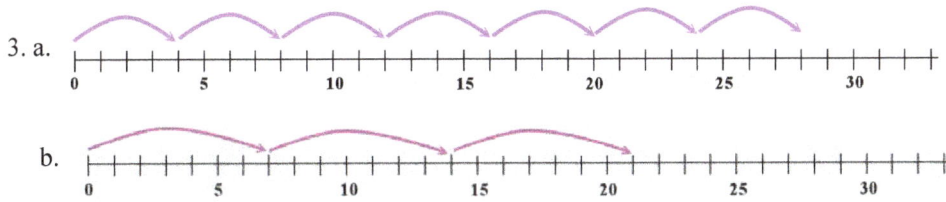

   b.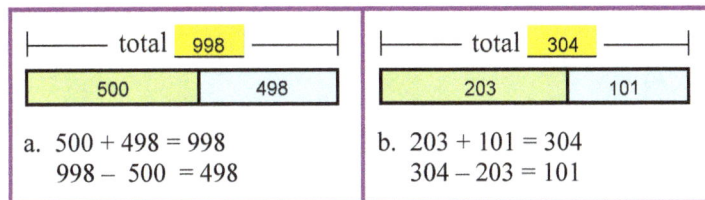

4.

— total 998 —	— total 304 —
500	498	203	101

   a. $500 + 498 = 998$ b. $203 + 101 = 304$
       $998 - 500 = 498$ $304 - 203 = 101$

Page 223

5. a. 72, 7, 54   b. 48, 35, 28   c. 36, 81, 72   d. 84, 64, 0

6. a. 6 hours   b. 5 hours
   c. 47 minutes   d. 25 minutes

7. a. Estimates will vary. For example: about 80 are blue.
   b. There are 85 blue ribbons. This is reasonable because it is close to the estimate.

8. 18 February

9. They are 21 days apart; three weeks apart.

10. a. 5:15   b. 11:45   c. 8:15   d. 12:45

Revision Chapter 6, p. 224

Page 224

1. a. $12.50   b. $80.05
   c. $6.50   d. $9.55   e. $1.40

2. a. $0.25   b. $0.90   c. 240 cents

3. a. The total cost is $3.55.   b. My change is $1.45.

4. a. She still needs to save $19.35.
   b. His total bill is $12.50. His change is $2.50.

Math Mammoth
Grade 3-B
Answer Key
International Version

By Maria Miller

Contents

Chapter 7: Four-Digit Numbers

Thousands, pp. 14-17

Page 14

1. See the chapter introduction.

Page 15

2. a. 1312   b. 1130
   c. 1057   d. 1502
   e. 2330   f. 3478

Page 16

3. a. 1256   b. 3594   c. 4617
   d. 9822   e. 6211   f. 5799

4. a. 1001   b. 2005   c. 4061
   d. 3012   e. 6200   f. 5090
   g. 1103   h. 7506   i. 5800
   j. 2011   k. 2320   l. 9032

Page 17

5. The numbers for the number lines are:

1096 1097 <u>1098</u> <u>1099</u> <u>1100</u> <u>1101</u>

<u>3406</u> 3407 3408 <u>3409</u> <u>3410</u> <u>3411</u>

<u>7329</u> <u>7330</u> <u>7331</u> <u>7332</u> <u>7333</u> 7334

6.

4010	4020	4030	4040	4050
4060	4070	4080	4090	4100
4110	4120	4130	4140	4150

7.

8610	8620	8630	8640	8650
8660	8670	8680	8690	8700
8710	8720	8730	8740	8750

Four-Digit Numbers and Place Value, pp. 18-22

Page 18

1.

thousands	hundreds	tens	ones
5	2	5	9

a. 5000 + <u>200</u> + 50 + <u>9</u>

thousands	hundreds	tens	ones
7	0	8	2

b. 7000 + 0 + <u>80</u> + <u>2</u>

Page 19

2.

a. 1034 = <u>1</u> thousand <u>0</u> hundreds <u>3</u> tens <u>4</u> ones
   = 1000 + <u>0</u> + <u>30</u> + 4

b. 5670 = <u>5</u> thousands <u>6</u> hundreds <u>7</u> tens <u>0</u> ones
   = 5000 + 600 + 70 + 0

c. 3508 = <u>3</u> thousands <u>5</u> hundreds <u>0</u> tens <u>8</u> ones
   = 3000 + 500 + 0 + 8

d. 8389 = <u>8</u> thousands <u>3</u> hundreds <u>8</u> tens <u>9</u> ones
   = 8000 + 300 + 80 + 9

e. 9007 = <u>9</u> thousands <u>0</u> hundreds <u>0</u> tens <u>7</u> ones
   = 9000 + 0 + 0 + 7

f. 6050 = <u>6</u> thousands <u>0</u> hundreds <u>5</u> tens <u>0</u> ones
   = 6000 + 0 + 50 + 0

g. 216 = <u>2</u> hundreds <u>1</u> ten <u>6</u> ones
   = 200 + 10 + 6

Four-Digit Numbers and Place Value, cont.

3.

a. Five thousand nine hundred and ninety	b. Six thousand and sixteen	c. Six thousand three hundred and three
T H T O 5 9 9 0	T H T O 6 0 1 6	T H T O 6 3 0 3
d. Eight thousand seven hundred	e. Nine thousand two hundred and forty-five	f. Ten thousand
T H T O 8 7 0 0	T H T O 9 2 4 5	ten thou-sands T H T O 1 0 0 0 0

4. a. 2090; 3200   b. 8005; 1087
   c. 8220; 2598   d. 4050; 2807

5. a. 3   b. 600
   c. 50   d. 2000

6. a. _6048_ , 6049, _6050_   b. _2323_ , 2324, _2325_
   c. _1799_ , 1800, _1801_   d. _8808_ , 8809, _8810_
   e. _7384_ , 7385, _7386_   f. _9243_ , 9244, _9245_

7. a. 7808   b. 9500
   c. 3004   d. 5060
   e. 4907   f. 8086
   g. 2047   h. 4620

8. a. 8000   b. 6
   c. 500   d. 40
   e. 200   f. 7000

9. a. 2390   b. 2420   c. 2445   d. 2485

10.

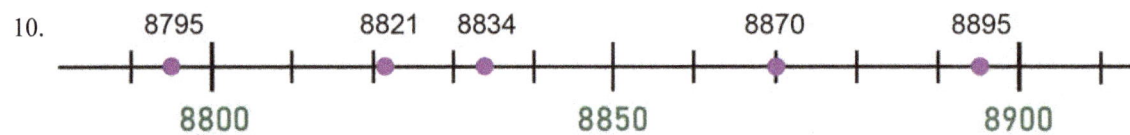

11. a. 8900   b. 9250   c. 9400   d. 9850

12.

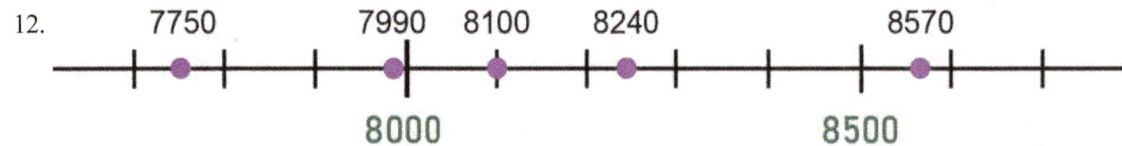

74

Four-Digit Numbers and Place Value, cont.

13. The number in the middle is 5826.

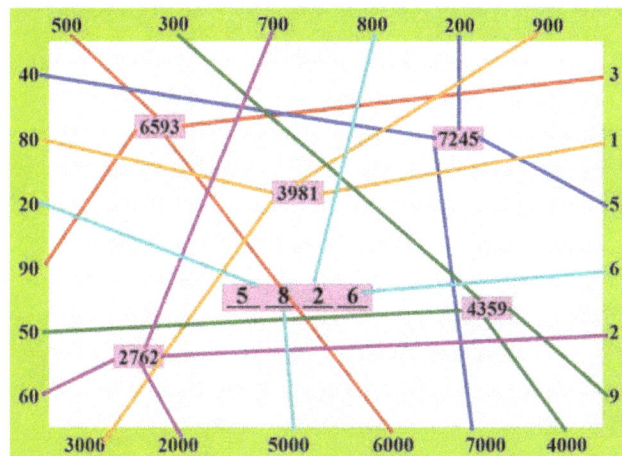

Puzzle Corner:

$$6 \times 8 + 5 \quad 2 \quad 1 = 569 \qquad 1 \quad 2 \times 6 + 5 \times 8 = 112$$

Comparing Numbers, pp. 23-24

1. a. 8502   b. 5710   c. 3811   d. 5743
   e. 5100   f. 6101   g. 9834   h. 9603

2.

a.	b.	c.	d.
1050 < 5095	220 < 1020	1307 > 1032	4012 < 4284
2400 < 2750	8060 > 6999	4906 < 6029	5008 < 5040
6005 > 4500	1007 < 1705	5077 < 5570	1890 < 1897

3. a. 7550   b. 2338
   c. 7099   d. 1212
   e. 1809   f. 3489

4. a. >   b. =
   c. <   d. <
   e. >   f. >
   g. =

5.

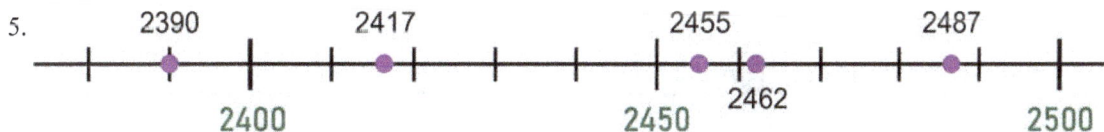

6. 3040 < 3899 < 4003 < 4203 < 4330

75

Add and Subtract Four-Digit Numbers 1, pp. 25-26

1. 2700, 2800, <u>2900</u>, <u>3000</u>, <u>3100</u>, <u>3200</u>

2.

a. 42 hundreds = 4200
   30 hundreds = 3000
	b. 56 hundreds = 5600
	   80 hundreds = 8000

3.

a. 600 + 400 = <u>*1000*</u>
   2500 + 500 = <u>*3000*</u>
   8200 + 800 = 9000
	b. 6600 + 400 = 7000
	   9300 + 700 = 10 000
	   1400 + 600 = 2000

4.

a. 5000 + 200 = 5200
   5100 + 400 = 5500
	b. 2900 + 200 = 3100
	   3100 + 300 = 3400

c. 6800 + 400 = 7200
   3800 + 800 = 4600
	d. 5600 − 200 = 5400
	   4500 − 300 = 4200

e. 8000 − 200 = 7800
   2200 − 600 = 1600
	f. 7900 − 800 = 7100
	   8500 − 700 = 7800

5.

a. 5000 + 1200 = 6200
   5100 + 2400 = 7500
	b. 2700 + 3200 = 5900
	   3100 + 6300 = 9400

c. 5000 − 2100 = 2900
   6000 − 3500 = 2500
	d. 8600 − 2500 = 6100
	   4500 − 3600 = 900

6. His trip was 3500 km one way and 7000 km both ways.

7. $5000 − $3700 = $1300; She still needs $1300.

8. a. △ = 300   b. △ = 500   c. △ = 500

   d. $x = 6500$ e. $y = 6900$ f. $z = 7400$

9.

a. 5000 − 800 = <u>4200</u>
   5000 − 1200 = <u>3800</u>
   5000 − 1700 = <u>3300</u>
   5000 − 2300 = <u>2700</u>
	b. 10 000 − 900 = <u>9100</u>
	   10 000 − 1700 = <u>8300</u>
	   10 000 − 2400 = <u>7600</u>
	   10 000 − 3500 = <u>6500</u>

Add and Subtract Four-Digit Numbers 2, pp. 27-28

1. a. 4000, 4010, <u>4020</u>, <u>4030</u>, <u>4040</u>, <u>4050</u>
   b. <u>1720</u>, <u>1730</u>, 1740, 1750, <u>1760</u>, <u>1770</u>
   c. <u>3350</u>, <u>3360</u>, 3370, 3380, <u>3390</u>, <u>3400</u>

Add and Subtract Four-Digit Numbers 2, cont.

2.

a. $100 + 20 = 120$ $5100 + 20 = 5120$	b. $220 + 40 = 260$ $4220 + 40 = 4260$
c. $140 - 90 = 50$ $4140 - 90 = 4050$	d. $230 - 30 = 200$ $4230 - 30 = 4200$

3. a. 5590 b. 7300
   c. 7730 d. 1440
   e. 2170 f. 5000

4. a. 6110   b. 4590   c. 8230
   d. 4060   e. 9700   f. 3500

5. a.

6140	6180	6220	6260	6300	6340	6380	6420

b.

2870	2920	2970	3020	3070	3120	3170	3220

c.

5450	5650	5850	6050	6250	6450	6650	6850

d.

1470	1500	1530	1560	1590	1620	1650	1680

6. a. 20   b. 50
   c. 50   d. 60

Puzzle corner. The puzzle has MANY possible solutions.
Basically you just pick one number at will and start filling
the puzzle in, and if you run into a difficulty, you change
the number. This is just an example solution.

4550	−	14	+	24	=	4560
−		+		−		
0	+	30	+	20	=	50
+		−		+		
30	+	14	+	56	=	100
=		=		=		
4580		30		60		

Add Four-Digit Numbers in Columns, pp. 29-30

Teaching box: 8423;   8255;   7423

1. a. 5601 b. 7109 c. 7740

2. a. 7386   b. 4770   c. 6818
   d. 9472   e. 8162   f. 9277

3. a. 6293 b. 4668

4. a. The total is $4594.   b. The total cost is $2220.

Puzzle corner:

```
   3  9  5  2          2  9  8  1
 + 5  1  2  9        + 2  4  3  6
 ───────────        ────────────
   9  0  8  1          5  4  1  7
```

Subtract 4-Digit Numbers with Regrouping, pp. 31-33

Page 31

Teaching box: 2895
Check: 2895 + 2244 = 5139

1. a. 4581   Check: 4581 + 510 = 5091
   b. 1197   Check: 1197 + 1716 = 2913
   c. 7024   Check: 7024 + 1378 = 8402
   d. 5970   Check: 5970 + 911 = 6881
   e. 3056   Check: 3056 + 3490 = 6546
   f. 4055   Check: 4055 + 5025 = 9080
   g. 3393   Check: 3393 + 1116 = 4509
   h. 4144   Check: 4144 + 2065 = 6209

Page 32

Teaching box: 5349

2. a. 1786   b. 2276
   c. 295   d. 6099
   e. 2510   f. 4926
   g. 1899   h. 6221

Page 33

3. a. 3285
   b. 4957

4. An equation is optional for this question, as it asks to solve it with mental maths.
   a. $4200 - s = 3100$. The track is 1100 metres shorter when using the shortcut
   b. $3100 + 3100 + 3100 = t$. He jogged 9300 metres.

Puzzle corner:

4 × 5 + 3 2 1 = 341

1 2 × 5 + 3 × 4 = 72

More Practice, pp. 34-36

Page 34

1. a. 5374   b. 4885
   c. 1133   d. 6289
   e. 4688   f. 1518
   g. 3186   h. 1737

2. △ = 6507

Page 35

3. 6207

4. The population after the changes is $2387 + 489 + 517 - 578 = \underline{2815}$.

5. You can find out that the number of fish grew by 250 each year. You can find out that in two years, the number of fish grew by 500.

6. a. $4000 - 1700 = l$ or $1700 + l = 4000$; $l = \$2300$
   b. $p - 500 - 700 = 1000$; $p = \$2200$

Page 36

7.
a.

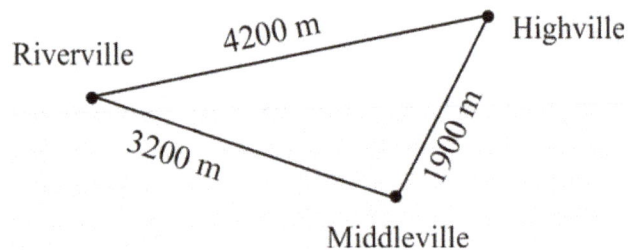

Riverville — 4200 m — Highville
3200 m
1900 m
Middleville

b. 4200 m + 1900 m + 3200 m = 9300 m;
   The total distance is 9300 m.
c. 3200 m + 3200 m = 6400 m.
   1900 m + 1900 m = 3800 m.
   6400 m – 3800 m = 2600 m.
   It is 2600 m further.

Page 36

Puzzle corner:

3	8	5	3
– 2	1	8	9
1	6	6	4

8	9	5	4
– 4	2	3	6
4	7	1	8

6	0	0	9
– 3	2	2	5
2	7	8	4

7	0	0	3
– 2	8	6	7
4	1	3	6

Word Problems, pp. 37-38

1. Student equations will vary.

   a. Using two equations: Total cost is C = $1158 + $745. C = $1903. $1903 + x = $2000; x = $97.
      Her change was <u>$97</u>. This can be written as one equation like this: y = $2000 – ($1158 + $745).

   b. B = $1109 + $1109 + $1109 + $1109 – $500. B = 3936. The total bill is <u>$3936</u>.

   c. P = 1380 – 704; P = 676. You have <u>676</u> pages left to read.

2. a. Yes, you can. The total cost is $979 + $979 + $979 = $2937. Change is $3000 – $2937 = <u>$63</u>.

   b. The largest number you can build is 4321 and the smallest is 1234. The difference between them is
      4321 – 1234 = <u>3087</u>.

   c. $8740 – $1295 = $7445. The price difference is <u>$7445</u>.

Puzzle corner:

```
  5 [2][3] 7        [6][3] 0  1        6 [7][6]9         [2] 1  8  8
– [1] 5  5 [3]     –  2  7 [5][9]     + [3] 0  0 [5]     +  2  7 [1][2]
 ─────────────     ─────────────     ─────────────     ─────────────
  3  6  8  4         3  5  4  2         9  7  7  4         4  9  0  0
```

Mixed Revision Chapter 7, pp. 39-40

1. a. 49   b. 65   c. 5
    249   665   275
    419   333   895

2. a. L = 6 × 2 + 4 × 4; L = 28. They have 28 legs.
   b. W = 9 × 4 + 3; W = 39. There are 39
      windows in total.

3. a. 570   b. 900   c. 600   d. 60
    340   260   430   1000

4. a. 80   b. 24   c. 110

5. 46 minutes

6. a. 4:15   b. 11:20   c. 10:50   d. 8:40

7. a. 4   b. 7   c. 8   d. 4
    9   5   6   8
    7   8   4   5

8. a. 8 × $5 = $40   You can buy 8 pairs of socks.
   b. 4 × 7 = 28   There will be 4 layers.

9. a. $10.40   b. $29.30   c. $43.60

Revision Chapter 7, pp. 41-42

1. a. 7240   b. 6005   c. 2029

2. a. 7503   b. 1037
    3090   6400

3. a. >   b. >
   c. <   d. <

4. a. 1900   b. 3300
    7200   3700

   c. 800   d. 4900
    1600   8300

5. a. 2334   b. 7089

6. 6370

7. a. △ = 700   b. △ = 9800   c. △ = 1500

8. a. 1669   b. 3208

9. a. $2863
   b. $3472

Chapter 8: Division

Division as Making Groups, pp. 47-49

<u>Page 47</u>

1.

a. There are <u>15</u> carrots. Make groups of 5. How many groups? 3 How many 5's are there in <u>15</u>? 3	b. There are 20 berries. Make groups of 4. How many groups? 5 How many 4's are there in 20? 5	c. There are 9 apples. Make groups of 3. How many groups? 3 How many 3's are there in 9? 3
d. There are 10 fish. Make groups of 2. How many groups? 5 How many 2's are there in 10? 5	e. There are 12 daisies. Make groups of 6. How many groups? 2 How many 6's are there in 12? 2	f. There are 16 camels. Make groups of 4. How many groups? 4 How many 4's are there in 16? 4

<u>Page 48</u>

2. a. $10 \div 2 = 5$. b. $20 \div 4 = 5$.
   c. $18 \div 6 = 3$. d. $9 \div 3 = 3$.
   e. $15 \div 5 = 3$. f. $21 \div 3 = 7$.

<u>Page 49</u>

3.

a. $18 \div 3 = 6$	b. $24 \div 2 = 12$
c. $21 \div 3 = 7$	d. $25 \div 5 = 5$
e. $15 \div 5 = 3$	f. $24 \div 8 = 3$

4. a. $20 \div 4 = 5$ b. $20 \div 2 = 10$ c. $30 \div 6 = 5$ d. $24 \div 3 = 8$
   e. $20 \div 5 = 4$ f. $21 \div 7 = 3$ g. $24 \div 6 = 4$ h. $20 \div 10 = 2$

Page 50

1. a. Two *groups of 6* is 12. $2 \times 6 = 12$ 12 divided into *groups of 6* is two groups. $12 \div 6 = 2$	b. Five **groups of 2** is <u>10</u>. $5 \times 2 = 10$ 10 divided into *groups of 2* is 5 groups. $10 \div 2 = 5$
c. One **group of 4** is 4. $1 \times 4 = 4$ 4 divided into *a group* *of 4* is one group. $4 \div 4 = 1$	d. Five *groups of 1* is 5. $5 \times 1 = 5$ 5 divided into *groups* *of 1* is 5 groups. $5 \div 1 = 5$
e. 7 *groups of 2* is 14. $7 \times 2 = 14$ 14 divided into *groups of 2* is 7 groups. $14 \div 2 = 7$	f. 6 *groups of 3* is 18. $6 \times 3 = 18$ 18 divided into *groups of 3* is 6 groups. $18 \div 3 = 6$

Page 51

2.

a. $2 \times 4 = 8$ $8 \div 4 = 2$	b. $6 \times 2 = 12$ $12 \div 2 = 6$
c. $4 \times 4 = 16$ $16 \div 4 = 4$	d. $2 \times 6 = 12$ $12 \div 6 = 2$
e. $1 \times 4 = 4$ $4 \div 4 = 1$	f. $2 \times 7 = 14$ $14 \div 7 = 2$

g. $3 \times 6 = 18$ $18 \div 6 = 3$	h. $4 \times 2 = 8$ $8 \div 2 = 4$	i. $1 \times 5 = 5$ $5 \div 5 = 1$

3.

a. $3 \times 5 = 15$ $15 \div 5 = 3$	b. $3 \times 8 = 24$ $24 \div 8 = 3$	c. $6 \times 5 = 30$ $30 \div 5 = 6$

Page 52

4.

a. $14 \div 2 = 7$ $7 \times 2 = 14$	b. $18 \div 2 = 9$ $9 \times 2 = 18$	c. $21 \div 7 = 3$ $3 \times 7 = 21$
d. $54 \div 6 = 9$ $9 \times 6 = 54$	e. $24 \div 4 = 6$ $6 \times 4 = 24$	f. $30 \div 3 = 10$ $10 \times 3 = 30$
g. $32 \div 4 = 8$	h. $56 \div 7 = 8$	i. $55 \div 5 = 11$

Division and Multiplication, cont.

5.

a.	b.	c.	d.
24 ÷ 4 = 6 16 ÷ 2 = 8 20 ÷ 2 = 10 36 ÷ 9 = 4	15 ÷ 5 = 3 35 ÷ 5 = 7 49 ÷ 7 = 7 54 ÷ 9 = 6	32 ÷ 8 = 4 40 ÷ 8 = 5 50 ÷ 5 = 10 42 ÷ 6 = 7	48 ÷ 6 = 8 56 ÷ 8 = 7 81 ÷ 9 = 9 100 ÷ 10 = 10

Puzzle corner. a. 10   b. 8   c. 50   d. 2   e. 1   f. 5

Multiplication and Division Fact Families, pp. 53-55

1. a. 4 × 6 = 24
       6 × 4 = 24
       24 ÷ 4 = 6
       24 ÷ 6 = 4

   b. 3 × 5 = 15
       5 × 3 = 15
       15 ÷ 5 = 3
       15 ÷ 3 = 5

   c. 7 × 4 = 28
       4 × 7 = 28
       28 ÷ 7 = 4
       28 ÷ 4 = 7

   d. 5 × 4 = 20
       4 × 5 = 20
       20 ÷ 5 = 4
       20 ÷ 4 = 5

2.
1 × 5 = 5
5 × 1 = 5
5 ÷ 5 = 1
5 ÷ 1 = 5

3. Student illustrations will vary but should show a 3 by 6 or 6 by 3 array.

3 × 6 = 18
6 × 3 = 18
18 ÷ 3 = 6
18 ÷ 6 = 3

4.

a. 7 × 5 = 35
    5 × 7 = 35
  35 ÷ 7 = 5
  35 ÷ 5 = 7
	b. 8 × 9 = 72
	    9 × 8 = 72
	  72 ÷ 8 = 9
	  72 ÷ 9 = 8
		c. 12 × 4 = 48
		    4 × 12 = 48
		  48 ÷ 4 = 12
		  48 ÷ 12 = 4

5. Seven rows.

6. I can find out how many trees there are in the orchard: 3 × 5 = 15 trees. I can find out how many mangos are in these trees, in total: 3 × 15 = 15 + 15 + 15 = 45. The trees have 45 mangos in total.

7. You can find out how many chairs were in each row. Since 5 × 9 = 45, there were 9 chairs in each row.

Multiplication and Division Fact Families, cont.

8. a. 9   b. 14   c. 5
   d. 8   e. 7   f. 35

9.

a.	b.	c.	d.
$18 \div 2 = \underline{9}$	$15 \div 3 = \underline{5}$	$40 \div 4 = \underline{10}$	$45 \div 5 = \underline{9}$
$16 \div 2 = \underline{8}$	$18 \div 3 = \underline{6}$	$16 \div 4 = \underline{4}$	$55 \div 5 = \underline{11}$
$24 \div 2 = \underline{12}$	$21 \div 3 = \underline{7}$	$36 \div 4 = \underline{9}$	$60 \div 5 = \underline{12}$

10.

a. Division table of six	b. Division table of seven	c. Division table of eight
$6 \div 6 = 1$	$7 \div 7 = 1$	$8 \div 8 = 1$
$12 \div 6 = 2$	$14 \div 7 = 2$	$16 \div 8 = 2$
$18 \div 6 = 3$	$21 \div 7 = 3$	$24 \div 8 = 3$
$24 \div 6 = 4$	$28 \div 7 = 4$	$32 \div 8 = 4$
$30 \div 6 = 5$	$35 \div 7 = 5$	$40 \div 8 = 5$
$36 \div 6 = 6$	$42 \div 7 = 6$	$48 \div 8 = 6$
$42 \div 6 = 7$	$49 \div 7 = 7$	$56 \div 8 = 7$
$48 \div 6 = 8$	$56 \div 7 = 8$	$64 \div 8 = 8$
$54 \div 6 = 9$	$63 \div 7 = 9$	$72 \div 8 = 9$
$60 \div 6 = 10$	$70 \div 7 = 10$	$80 \div 8 = 10$
$66 \div 6 = 11$	$77 \div 7 = 11$	$88 \div 8 = 11$
$72 \div 6 = 12$	$84 \div 7 = 12$	$96 \div 8 = 12$

Dividing Evenly into Groups, pp. 56-58

1. a. $12 \div 2 = 6$ b. $6 \div 2 = 3$ c. $10 \div 2 = 5$

2. a. $12 \div 3 = 4$ b. $6 \div 3 = 2$ c. $24 \div 3 = 8$

3. a. $8 \div 4 = 2$ b. $12 \div 4 = 3$ c. $20 \div 4 = 5$

4. a. $8 \div 2 = 4$ b. $10 \div 5 = 2$ c. $21 \div 3 = 7$

   d. $21 \div 1 = 21$ e. $30 \div 10 = 3$ f. $14 \div 2 = 7$

5. a. 5, 2, 8   b. 4, 10, 5   c. 4, 6, 8

6. See the chapter introduction.

7.

a. $18 \div 3 = 6$ They each got six marbles.	b. $4 \times 7 = 28$ There was a total of 28 marbles.
c. $24 \div 6 = 4$ The pieces were 4 metres long.	d. $24 \div 3 = 8$ Each girl got 8 hairpins.

8. a. and b. Answers will vary. Please check the student's work. For example:
   a. Mary shared 20 apples evenly between four horses. How many apples did each horse get?
   b. Six children played together, and they had 24 toy cars in the game. They shared them evenly.
      How many cars did each child get?

Multiplication and Division Word Problems, pp. 59-60

Page 59

1.

a. $90 \div 10 = 9$	b. $12 \times 8 = 96$
Nine pages are full of stamps.	She has 96 stamps.

Page 60

2.

a. $4 \times 11 = 44$	b. $12 \div 4 = 3$
There would be 44 children.	You would need three taxis.
c. $10 \times 5 = 50$	d. $10 \div 5 = 2$
There are 50 eggs.	He used 2 bags.
e. $3 \times 5 = 15$	f. $18 \div 3 = 6$
She can fit 15 bottles of juice.	She will need six bags.
g. $36 \div 3 = 12$	h. $25 \div 5 = 5$
Each one got 12 cherries.	Each group had five students.
i. $7 \times 5 = 35$	j. $72 \div 6 = 12$
There were 35 people.	Each part was 12 centimetres long.

More Word Problems, pp. 61-63

Page 61

1. $(14 + 13) \div 3 = c$; $c = 9$. Each person got 9 cherries.

2. a. $C = (4 + 6 + 7 + 5) \div 2$. $C = 11$. Mum used 11 containers.
   b. $C = 4 \times 12 + 7$. $C = 55$. There are 55 crayons.
   c. $x = 7 \times 10 + 8$. $x = 78$. There are 78 chairs.

Page 62

3.

a. $20 \div 2 = 10$ or $2 \times 10 = 20$	b. $3 \times 7 = 21$
She filled ten jars.	She spent 21 hours in total.
c. $24 \div 8 = 3$ or $3 \times 8 = 24$	d. $60 \div 12 = 5$ or $5 \times 12 = 60$
She can make three omelettes.	You will need five boxes.

4. a. You can find out how many vans are needed. $20 \div 5 = 4$ or $4 \times 5 = 20$. Four vans are needed.
   b. You can find out how many hairpins Erica has in total. $4 \times 20 = 80$. She has 80 pins.
   c. You can find out how many small poster boards Brian needs to make the two big ones. $2 \times 4 = 8$.
      He needs 8 small poster boards.
   d. You can find out how many marbles are in each row. $30 \div 5 = 6$. There are six marbles in each row.

Zero in Division, pp. 63-65

Page 63

1.

a. $4 \div 1 = 4$	b. $14 \div 14 = 1$	c. $1 \div 1 = 1$	d. $0 \div 5 = 0$
$4 \div 0 = $ ___	$0 \div 0 = $ ___	$7 \div 0 = $ ___	$5 \div 5 = 1$
e. $0 \div 1 = 0$	f. $0 \div 14 = 0$	g. $0 \div 3 = 0$	h. $10 \div 10 = 1$
$0 \div 4 = 0$	$14 \div 0 = $ ___	$0 \div 1 = 0$	$1 \div 1 = 1$

Zero in Division, pp. 63-65

2.

a. $6 \times 1 = 6$ $6 \div 1 = 6$	b. $0 \times 8 = 0$ $0 \div 8 = 0$	c. $5 \times 7 = 35$ $35 \div 7 = 5$
d. $10 \times 11 = 110$ $110 \div 11 = 10$	e. $1 \times 1 = 1$ $1 \div 1 = 1$	f. $1 \times 8 = 8$ $8 \div 8 = 1$
g. $0 \times 0 = 0$ not possible	h. $5 \times 9 = 45$ $45 \div 9 = 5$	i. $9 \times 0 = 0$ not possible

3. a. How many cars did each boy get?

$18 + 7 + 11 = 36$. $36 \div 3 = 12$. Each boy got 12 cars.

b. How much did she pay them in total?

$5 \times 10 + 15 = \$65$. She paid them \$65 in total.

Puzzle corner:

a. no solutions

b. Any number is a solution (there are an infinite number of solutions)

c. & d. No solutions.

Division Practice, pp. 65-67

1.

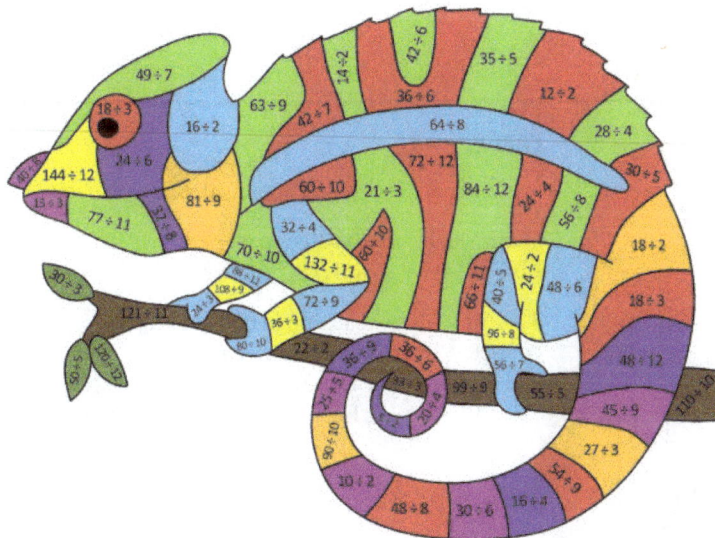

2. See the chapter introduction.

3. a. How many days will it take him to read the books?

$32 + 40 = 72$. $72 \div 12 = 6$.

b. How many pieces will Kelly get?

$80 \div 20 = 4$ and $40 \div 20 = 2$. $4 + 2 = 6$. She gets 6 pieces.

4. a. $9 \div 1 = 9$ ~~$9 \div 0 =$~~	b. $0 \div 20 = 0$ ~~$20 \div 0 =$~~	c. $11 \div 1 = 11$ ~~$8 \div 0 =$~~	d. ~~$0 \div 0$~~ $0 \div 10 = 0$

5. a. $30 \div 6 = 5$ Each bag had 5 kg.	b. $4 \div 4 = 1$ Each glass had one cup of milk.
c. $6 \times 7 = 42$ They can take 42 passengers.	d. $56 \div 7 = 8$ You need eight taxis.

6. Student equations will vary. For example: $3 \times 12 + E = 41$, or $36 + E = 41$ or $E = 41 - 36$. $E = 5$.
The carton that was not full had 5 eggs.

7. a. 8, 6, 7   b. 8, 12, 9   c. 45, 81, 99   d. 7, 10, 4

Division Practice, cont.

8.

a. 20 ÷ 2 = 10 22 ÷ 2 = 11 24 ÷ 2 = 12 26 ÷ 2 = 13 28 ÷ 2 = 14 30 ÷ 2 = 15 32 ÷ 2 = 16 34 ÷ 2 = 17	b. 40 ÷ 20 = 2 80 ÷ 20 = 4 120 ÷ 20 = 6 160 ÷ 20 = 8 200 ÷ 20 = 10 240 ÷ 20 = 12 280 ÷ 20 = 14 320 ÷ 20 = 16	c. 45 ÷ 5 = 9 55 ÷ 5 = 11 65 ÷ 5 = 13 75 ÷ 5 = 15 85 ÷ 5 = 17 95 ÷ 5 = 19 105 ÷ 5 = 21 115 ÷ 5 = 23

9. a. 4   b. 50   c. 6
   d. 49   e. 8   f. 40

Puzzle corner: There are many possible answers. These are just some examples.

20	÷	4	= 5
÷		÷	
5	÷	1	= 5
=		=	
4		4	

54	÷	9	= 6
÷		÷	
6	÷	3	= 2
=		=	
9		3	

Missing Numbers, pp. 68-69

1. a. 6   b. 4   c. 20   d. 70

2. a. 4   b. 7   c. 7   d. 9
   e. 9   f. 7   g. 8   h. 48

3. a. 28   b. 9   c. 12   d. 121

4.

a. 12 ÷ 2 = 6 6 × 2 = 12	b. 22 ÷ 2 = 11 11 × 2 = 22	c. 16 ÷ 2 = 8 or 16 ÷ 8 = 2 8 × 2 = 16
d. 24 ÷ 3 = 8 8 × 3 = 24	e. 32 ÷ 4 = 8 8 × 4 = 32	f. 25 ÷ 5 = 5 5 × 5 = 25

5. a. 6, 8, 8   b. 11, 12, 9   c. 10, 8, 5

6. a. $x = 14$ b. $y = 30$ c. $s = 28$
   d. $v = 10$ e. $w = 7$ f. $z = 7$

7. b. 30 ÷ 5 = 6. Each plate had six grapes.
   c. 20 ÷ 4 = 5. He made five stacks.
   d. 3 × 20 = 60. Ken read 60 pages.
   e. 30 ÷ 3 = 10. Each child planted 10 plants.

Bar Graphs, pp. 70-73

1. a. Who read the most books? Emilia   How many? 18
   b. The child that read the most books read 10 more books than the child
      that read the least.
   c. How many more books did Grayson read than Owen? 3 books
   d. How many more books did Emilia read than Ava? 6 books
   e. Who read more, Nora and Ava together or Oliver and Finn together?
       Oliver and Finn . (They read 22 books together.)
   f. Altogether, Emilia, Violet, Nora, and Ava read 54 books.

Bar Graphs, cont.

Page 71

2.

Month	Dogs sold
March	15
April	25
May	20
June	30
July	20
Total	110

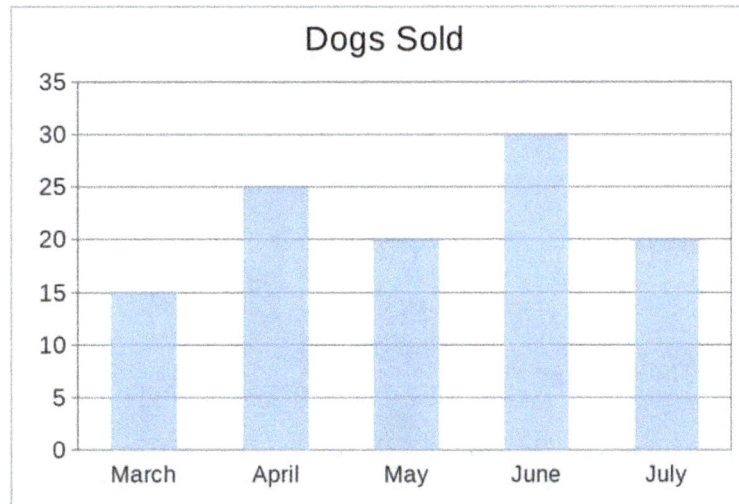

3. Graphs will vary; check the student's graph. A scale of 1 or 2 does not work for this graph. Since the largest number is 21, a scale of 3 works well. But a student could choose some other scale, such as a scale of 4.

Name	Cans of cat food
Luna	15
Simba	11
Oliver	16
Leo	21
Total	63

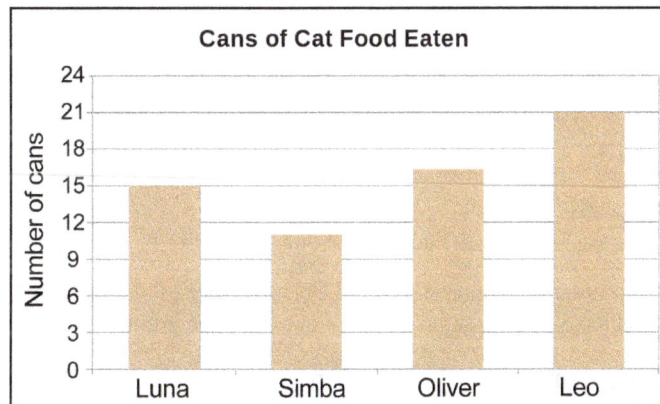

4. a. Luna ate _4_ more cans than Simba.
   b. Luna & Simba ate _5_ more cans than Leo.
   c. He ate _10_ more cans.
   d. This question is just to help the student learn to think. Check if their answer sounds reasonable.
      Examples: Leo is larger than the other cats, or Leo is more active than the other cats.

Page 72

5. Graphs will vary; check the student's graph. Since the largest number is 30, a scale of 4 works well. But a student could choose some other scale, such as a scale of 5.

Child	Seashells
Olivia	24
Emma	8
Sofia	30
Liam	14
Aiden	19
Total	95

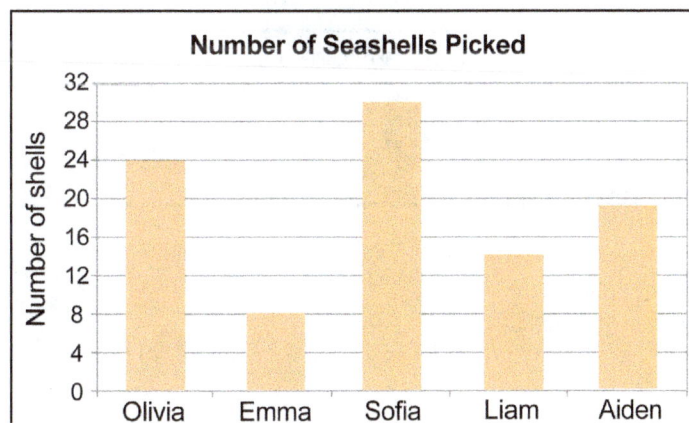

Bar Graphs, cont.

6. Graphs will vary; check the student's graph. Since the largest number is 90, a scale of 10 works well.

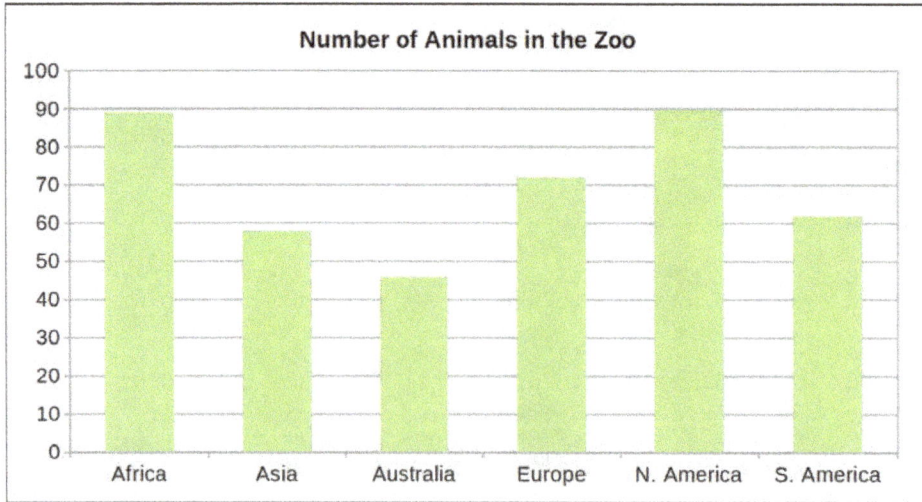

Number of Animals in the Zoo

	Africa	Asia	Australia	Europe	N. America	S. America

7.

Day	Mon	Tues	Wed	Thurs	Fri	Sat
Bottles of water	about 30	about 40	about 40	about 40	about 50	about 70

   a. about <u>40 more</u> bottles of water
   b. about <u>120</u> bottles of water on Friday and Saturday; about <u>150</u> bottles of water on the other four days

8. Graphs will vary; check the student's graph. Since the largest number is 100, a scale of 10 works well.
   The bars could also be two blocks wide instead of one. Check that the graph matches the table in the worktext.

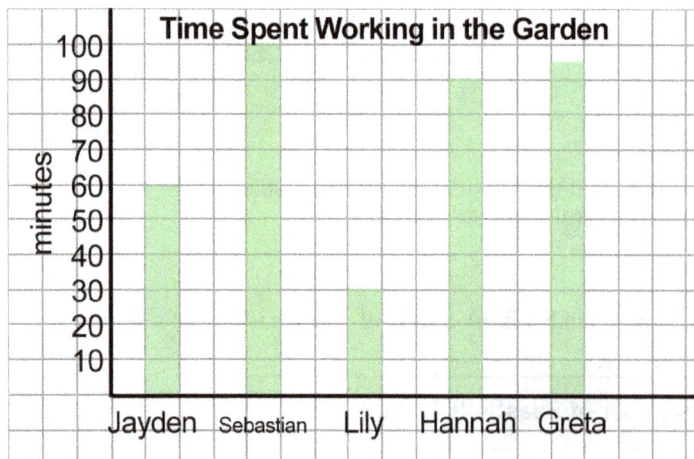

Time Spent Working in the Garden

minutes

Jayden	Sebastian	Lily	Hannah	Greta

Pictographs, pp. 74-76

1. a. Ava picked <u>16</u> more flowers.
   b. Sofia and Ava picked <u>40</u> together.
   c. Both Oliver & Liam and Sofia & Ava picked the same amount — 40 flowers, together.

Pictographs, cont.

Page 74

2.

	Fruits We Picked
= 10 oranges = 10 mangos = 10 bananas = 10 apples	**oranges** 🍊🍊🍊🍊
	mangos 🥭
	bananas 🍌🍌🍌🍌🍌
	apples 🍎🍎🍎

Page 75

3. Pictographs may vary, based on how many kilograms of vegetables the student chooses to be represented by one carrot. Now, 5 kg works well, since all of the amounts are evenly divisible by five. The student may decide to use 10, and draw half carrots to represent five. If they choose a smaller number than 5, encourage the student to think of a number they could use to save time by not having to draw as many carrots.

Vegetable use in one month	
Jacksons	🥕🥕🥕
Joneses	🥕🥕
Millers	🥕🥕🥕🥕
Eastmans	🥕🥕🥕🥕🥕🥕
Davises	🥕🥕🥕🥕🥕

🥕 = __5__ kg of vegetables

4. a. 15 kg more
   b. 5 kg more
   c. 75 kg total

5. Answers will vary. The pictograph below uses one picture to mean 4 packages. The student may use one picture to signify 2 packages.

Hall Family Packages	
Mum	📦📦📦
Dad	📦📦📦
Isabella	📦
Cayden	📦

Key: 📦 = 4 packages

Page 76

6. Answers will vary; check the student's pictograph.
   For example:

Milk	
January	🥛🥛🥛🥛🥛
February	🥛🥛🥛🥛🥛🥛
March	🥛🥛🥛🥛🥛🥛
April	🥛🥛🥛🥛🥛🥛🥛
May	🥛🥛🥛🥛
June	🥛🥛🥛🥛🥛

Key: 🥛 = 2 litres

7. In the first three months of the year, the family consumed 39 litres, and in the following three months they consumed 36 litres. So, in the first three months they consumed <u>3 more litres</u> than in the next three months.

8. a. Example pictograph:
   b. He got 400 kg more on Friday.
   c. He caught 3400 kilograms in total.

Fish Caught by Jack	
Monday	🐟🐟🐟
Wednesday	🐟🐟🐟🐟🐟🐟🐟🐟
Friday	🐟🐟🐟🐟🐟
Sunday	🐟🐟🐟

🐟 = 200 kg of fish

1.

a. Divide 14 bananas among 3 people. Each person gets _4_ bananas and _2_ bananas are left over.  **4 R2**	b. Divide 14 carrots among 3 people. Each person gets _4_ carrots and _2_ carrots are left over.  _4_ R _2_
c. Divide 8 pears among 5 people. Each person gets _1_ pear and _3_ pears are left over.  _1_ R _3_	d. Divide 14 apples among 4 people. Each person gets _3_ apples and _2_ apples are left over.  _3_ R _2_
e. Divide 15 hens into 6 boxes. Each box has _2_ hens, and _3_ hens are left over.  _2_ R _3_	f. Divide 9 fish between 2 people. Each person gets _4_ fish and _1_ fish is left over.  _4_ R _1_

2.

a. Divide into groups of 3.	b. Divide into groups of 4.	c. Divide into groups of 6.	d. Divide into groups of 5.
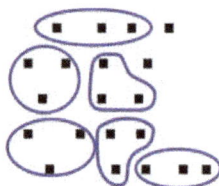 _6_ groups _2_ dot(s) left over	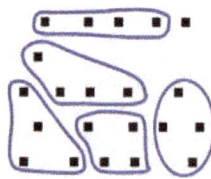 _5_ groups _1_ dot(s) left over	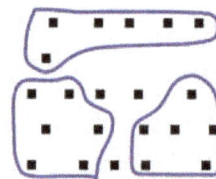 _3_ groups _3_ dot(s) left over	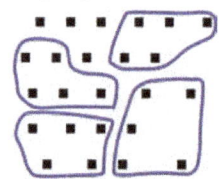 _4_ groups _4_ dot(s) left over

When Division Is Not Exact, cont.

3.

a. Make 3 groups. 2 groups of 7 1 group of 8	b. Make 3 groups. 1 group of 6 2 groups of 7	c. Make 4 groups. 2 groups of 3 2 groups of 4

4. a. You will get four vases. b. Each family gets five bottles, and two are left over.
c. Four cartons are full.   d. They could each get four and split the remaining cookie in half.
e. Each got three rolls. Five were left.   f. Make three groups of eight and one group of seven.

Mixed Revision Chapter 8, pp. 80-81

1. a. 3, 8, 5   b. 3, 11, 2   c. 7, 8, 9   d. 5, 9, 7

2. a. 293   b. 466   c. 486   d. 162

3.

a. $99 + \boxed{46} = 145$ $145 - 99 = \boxed{46}$	b. $34 + \boxed{42} = 76$ $76 - 34 = \boxed{42}$

4. a. 5   b. 7   c. 174

5. a. <   b. >
c. <   d. >
e. <

6.

7. a. 56 minutes   b. 24 minutes   c. 4 minutes

8. There are several ways to make these amounts of change with coins and banknotes. Please check that the student's
drawings have the correct amounts.
a. Change: $4.10   b. Change: $18.60

9. a. $10 - (40 - 30) = 0$ b. $(4 + 5) \times 2 - 1 = 17$ c. $5 \times (7 - 3) - 1 = 19$
d. $5 + 10 \times (4 - 3) = 15$ e. $50 - (15 + 2) - 9 = 24$ f. $6 + (7 - 3) \times 2 = 14$

Page 82

1. a. $2 \times 6 = 12$ b. $3 \times 5 = 15$
    $12 \div 6 = 2$ $15 \div 5 = 3$

2.

a.	b.	c.	d.
$36 \div 6 = 6$	$36 \div 3 = 12$	$56 \div 7 = 8$	$0 \div 9 = 0$
$3 \div 3 = 1$	$60 \div 6 = 10$	$72 \div 9 = 8$	$16 \div 16 = 1$
$4 \div 1 = 4$	$54 \div 9 = 6$	$100 \div 10 = 10$	$12 \div 1 = 12$

3.

a.	b.	c.
$7 \times 6 = 42$	$8 \times 1 = 8$	$7 \times 7 = 49$
$6 \times 7 = 42$	$1 \times 8 = 8$	$7 \times 7 = 49$
$42 \div 6 = 7$	$8 \div 8 = 1$	$49 \div 7 = 7$
$42 \div 7 = 6$	$8 \div 1 = 8$	$49 \div 7 = 7$

4. a. 6   b. 20   c. 9   d. 9

5.

a. $6 \times 0 = 0$	b. $1 \times 9 = 9$	c. $0 \times 0 = 0$
$0 \div 6 = 0$	$9 \div 1 = 9$	~~$0 \div 0$~~
~~$6 \div 0$~~	$9 \div 9 = 1$	

Page 83

6. a. $6 \times 8 = 48$ She has 48 crayons.   b. $24 \div 6 = 4$ There were four groups of six children.
   c. $48 \div 6 = 8$ She had eight bags of cookies.   d. $4 \times 3 = 12$ or $12 \div 3 = 4$. She wrote 4 invitations.

7. a. Student graphs will vary since the scaling on the vertical axis is chosen by the student.

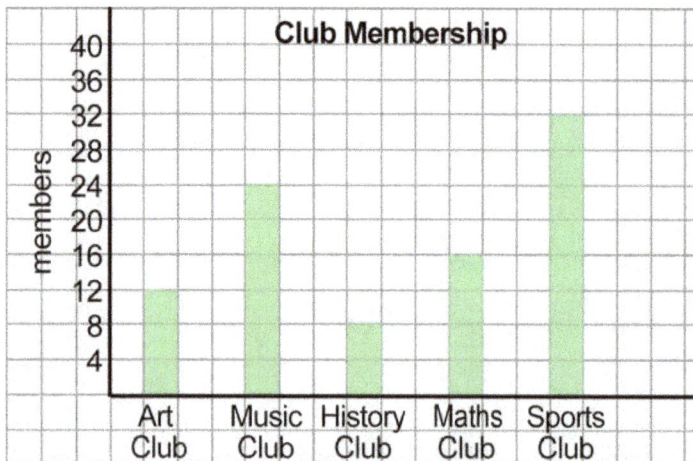

   b. There are 16 more students in the sports club than in the maths club.
   c. There are 68 students in total.

Chapter 9: Measuring

Centimetres and Millimetres, pp. 87-89

Page 87

1. a. 3 cm 4 mm = 34 mm
   b. 7 cm 7 mm = 77 mm

Page 88

1. c. 11 cm 6 mm = 116 mm
   d. 12 cm 9 mm = 129 mm
   e. 6 cm 1 mm = 61 mm
   f. 5 cm 3 mm = 53 mm

2. Check the student's answers. The answers below may not be the right length when printed from the download version, because many printers will print with "shrink to fit" or "fit to printable area."

   a. _____

   b. _____

   c. ____

   d. _____

   e. _____

Page 89

3. Answers will vary. Check the student's work.

4. a. 20 mm   b. 11 mm   c. 45 mm
   50 mm   18 mm   78 mm
   80 mm   23 mm   104 mm

5. The answers below may not match what you measure, if you have printed from the download version, because many printers will print with "shrink to fit" or "fit to printable area."
   side AB 53 or 54 mm
   side BC 110 mm
   side CA 117 mm

6. 280 or 281 mm

Line Plots, pp. 90-91

Page 90

1. a. Three pencils
   b. 8 1/2 cm
   c. 5 1/2 cm

2. Answers will vary. Please check the student's work.

Page 91

3. Answers will vary. Please check the student's work.

4. Answers will vary.

Grams and Kilograms, pp. 92-95

Page 92

1. Answers will vary.

Page 93

2. Answers will vary.

3. Answers will vary.

4. Answers will vary.

Page 94

5. a. 2 kg 200 g   b. 0 kg 200 g   c. 1 kg 400 g
   d. 0 kg 800 g   e. 3 kg 0g   f. 3 kg 400 g

6. a. The difference is 66 grams.
   b. The net weight is that of only the beans and water in the can. The additional 66 g is the weight of the can.
   c. 415 g + 415 g = 830 g. A third can makes it 830 g + 415 g = 1245 g. So she needs three cans to have at least 1 kg of beans.

7. Its mass is about 5 g.

Page 95

8. a. 5 g   b. 70 kg   c. 1 kg   d. 1 kg
   e. 30 g   f. 35 g   g. 2000 kg   h. 100 g
   i. 10 kg   j. 10 g   k. 5 kg   l. 300 g

9. An adult woman - 55 kg
   A puppy - 1 kg
   A pencil - 50 g
   A school book - 500 g
   A magazine - 150 g
   A 9-year-old boy - 25 kg

10. a. 3 kg
    b. 300 g
    c. 30 kg
    d. 300 g
    e. 3 kg
    f. 60 kg
    g. 250 g

Millilitres and Litres, pp. 96-97

Pages 96

1. Answers will vary.

Page 97

2. - 4. Student activities. Answers will vary.

5. a. 30 ml   b. 450 ml   c. 120 ml

Word Problems and More, pp. 98-99

Page 98

1. It contains 522 ml more of shampoo.

2. The baby's mass was 5 kg.

3. Eleven packages is only 990 g, so you will need twelve packages to have at least a kilogram.

4. a. The can is 70 grams.
   b. 330 g + 70 g = 400 g

5. a. a. 30 ml   b. 450 ml   c. 120 ml

Page 99

6. a. They contain 1350 ml of water.
   b. It is 350 ml more than 1 litre.

Page 99

7. You can fill four 250 ml glasses from one litre.

8. a. 4 × 150 g = 600 g; You need four apples.
   b. 7 × 150 g = 1050 g; You need seven apples.

9. 2 × 2 × 600 = 2400; She walked 2400 m.

10. She is 123 centimetres tall.

11. a. A 100-centimetre line will have twenty cars.
    b. It will take 60 cars to make a line 3 metres long.

12. Since each card weighed 50 g, there were 600 ÷ 50 = 12 people who attended her party.

Mixed Revision Chapter 9, pp. 100-102

Page 100

1. a. Estimate: $150 + $130 = $280;   Exact: $154 + $128 = $282.
   b. Estimate: $700 + $700 = $1400;   Exact: $698 + $698 = $1396.

2. a. 2777, 2778, 2779   b. 6059, 6060, 6061

   c. 7149, 7150, 7151   d. 6999, 7000, 7001

3.

a. 56 ÷ 7 = 8	b. 48 ÷ 6 = 8	c. 54 ÷ 9 = 6	d. 48 ÷ 8 = 6
49 ÷ 7 = 7	72 ÷ 6 = 12	81 ÷ 9 = 9	72 ÷ 8 = 9
28 ÷ 7 = 4	54 ÷ 6 = 9	36 ÷ 9 = 4	32 ÷ 8 = 4

4.

a. 6 × 7 = 42	b. 23 × 1 = 23	c. 8 × 9 = 72
42 ÷ 7 = 6	23 ÷ 1 = 23	72 ÷ 9 = 8
42 ÷ 6 = 7	23 ÷ 23 = 1	72 ÷ 8 = 9

Page 101

5. a. 4946 Check: 4946 + 2316 = 7262   b. 2761 Check: 2761 + 3242 = 6003

6. She will read 9 pages each day.

7. a. How many rows of cars did Diego have? 9 rows
   b. How much did Lucia earn last month? $27

8. The total cost is $787 + $787 + $1255 + $1255 = $4084.
   Check by adding the digits in columns in a different order.

9. a. 112   b. 91   c. 91   d. 62   e. 97   f. 51

Page 102

10.

a. 5 × 6 = 30	b. 6 × 7 = 42	c. 9 × 3 × 3 = 81
3 × 6 = 18	4 × 7 = 28	8 × 2 × 4 = 64
8 × 9 = 72	5 × 12 = 60	6 × 3 × 3 = 54
7 × 7 = 49	8 × 12 = 96	2 × 6 × 2 = 24

11. That is correct. From 6:20 to 6:30 is 10 minutes.

12. a. They have 30 stuffed animals in total.
    b. Mia has 9 more animals.
    c. 18 + 6 = 24; 24 ÷ 2 = 12; They would each get 12.

d. Stuffed animals

Mia	
Logan	
Zoe	
Mason	

= 3 stuffed animals

Page 103

1. a. _____

   b. _____

2. AB: _51_ mm
   BC: _72_ mm
   CA: _92_ mm

However, if you printed the lesson yourself, and didn't print at 100% but with "shrink to fit," "print to fit," or similar, the measurements will be smaller numbers than those given above. Please check the student's answers.
For example, the student might get:

AB: _47_ mm
BC: _68_ mm
CA: _86_ mm

3. Answers will vary.

4. Answers will vary.

Page 104

5. mm, cm, m, km

6. a. A butterfly's wings were 6 _cm_ wide.   b. Sherry is 86 _cm_ tall.
   c. Jessica jogged 5 _km_ yesterday.   d. The box was 60 _cm_ tall.
   e. The distance from the city   f. The room was 4 _m_ wide.
      to the little town is 80 _km_   g. The rubber is 3 _cm_ long

7. a. 800 g b. 3 kg c. 1 kg 400 g

8. a. Mum bought 5 _kg_ of apples.   b. Mary drank 350 _ml_ of juice.
   c. Dr. Smith weighs about 70 _kg_.   d. The banana weighed 150 g.
   e. The pan holds 2 _L_ of water.   f. A cell phone weighs about 100 g.

Page 105

9. a. _9_ cm _10_ cm _8_ cm _9 1/2_ cm _10_ cm
    _9 1/2_ cm _9_ cm _9_ cm _11_ cm _9 1/2_ cm
    _8 1/2_ cm _9 1/2_ cm

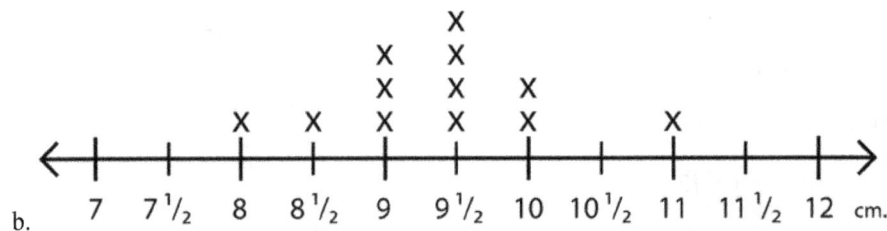

  b.

Chapter 10: Geometry

Polygons, pp. 111-115

Page 111

1. a. & b. Student's answers will vary. Be sure they give reasons for their sorting based on certain characteristics of the shapes, rather than just randomly arranging them. For example, they might sort them by the number of sides, by being a polygon or not, or by being a regular polygon or not. Or by some other criteria.

Page 113

2. Shapes 7, 11, and 13 are not polygons.

3. Shapes 3, 5, 6, and 10 are regular polygons.

4. Polygons 1, 2, 9, 10, 12, and 14 have four angles.

5. Polygons 4 and 5 have five sides.

Page 114

6. a. octagon   b. quadrilateral   c. pentagon
   d. not a polygon   e. octagon

Page 115

7. Answers will vary. The shapes should have five straight sides that all connect at angles. See examples on the right:

8. They are both regular polygons.
   3, 5, 6, and 10 from #1 are also regular polygons.

9. Answers will vary. These images show one example.

a. four triangles	b. a triangle and a pentagon	c. four quadrilaterals

10.

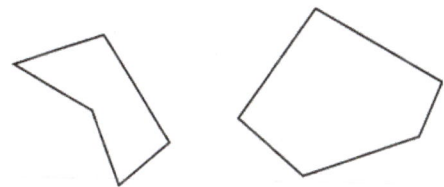

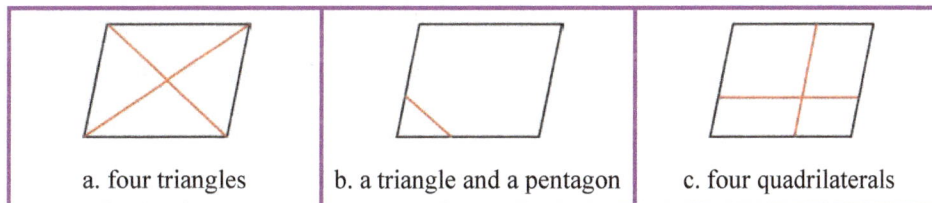

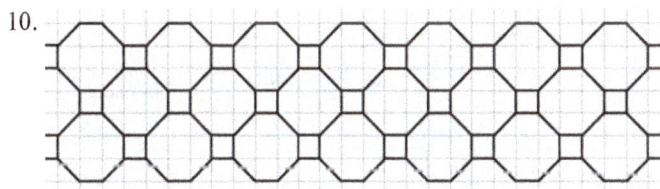

Some Special Quadrilaterals, pp. 116-117

Page 116

1. a. no   b. no   c. yes

2. a. yes   b. no   c. yes

Some Special Quadrilaterals, cont.

3. Answers will vary.

4. Yes. A rhombus with right angles is a square. Student drawings will vary.

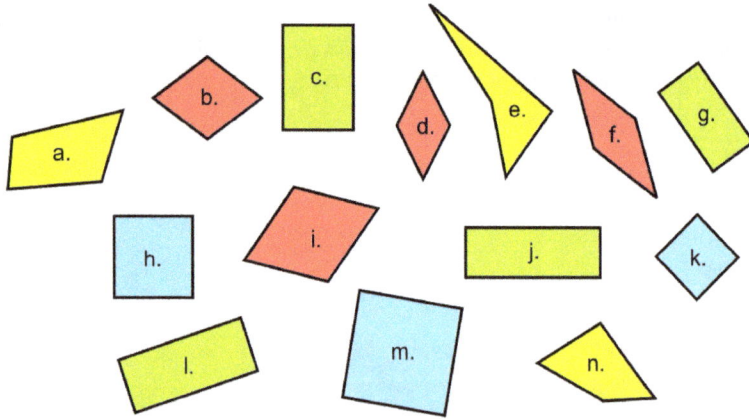

5.

The rectangles are c, g, j, and l. The squares are h, k, and m. The rhombi are b, d, f, and i. Other quadrilaterals are a, e, and n.

More Practice with Shapes, pp. 118-120

1. Macy is partly right, in that it *is* a square, but Brian is correct that it fits the characteristics of a rectangle.

2. a. Both shapes have at least three right angles.   b. Both shapes have four sides.   c. Both are regular polygons.

3. a. Shape 1, a regular pentagon
   b. Shape 3, a quadrilateral
   c. Shape 6, a square
   d. Shape 4, a rectangle
   e. Shape 2, a pentagon

4. a. & b. Drawings will vary. For example:

5. a., b. & c. Drawings will vary. For example:

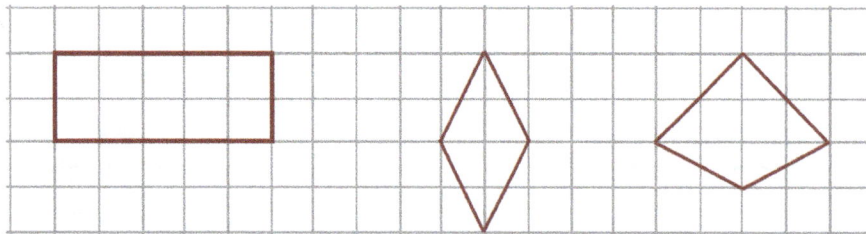

6. Student activity.

7. Drawings will vary. For example:

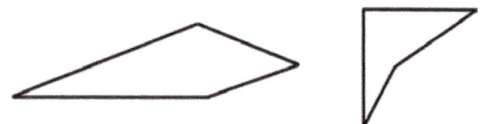

More Practice with Shapes, cont.

8. Students may use different colours or a different pattern of colouring.

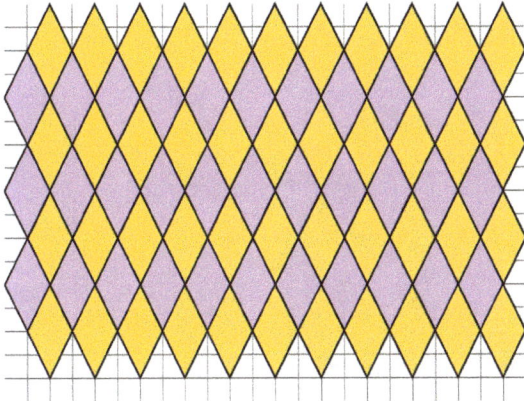

Puzzle Corner: Neither is correct. There are four.
Shapes 2, 3, 4, and 6 are all parallelograms.

Getting Started with Area, pp. 121-124

1. Shape 1 has the largest area. It is 18 square units, and shapes 2 and 3 are 16 square units each.

2. a. 8 square units (note that the centre of this figure is cut out)   b. 13 square units
   c. 8 square units   d. 12 square units

3. a. Rectangles can be 1 × 8 or 2 × 4.   b. Rectangles can be 1 × 15 or 3 × 5.   c. Rectangles must be 1 × 5.

4. a. & b. The rectangles can be 1 × 12, 2 × 6 or 3 × 4. Therefore, they can have 1, 2, 3, 4, 6, or 12 rows of little squares.

5. The rectangles can have:

Rows		Columns
1	×	12
2	×	6
3	×	4
4	×	3
6	×	2
12	×	1

6. That is not correct. The squares are not covering the figure evenly and there are gaps. It is actually 13 square units.

7. 21 square units

Puzzle Corner: 16 square units. You will notice that the tiles along the line on the left side of the figure are split in half, so each of those will count as 1/2 square unit. See the image on the right.

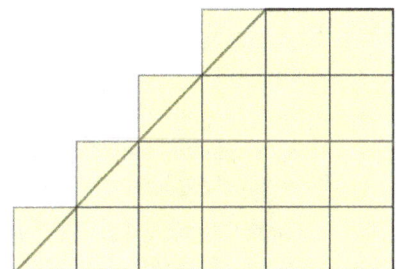

99

Units for Measuring Area, pp. 125-127

1. a. 9 cm^2 b. 22 cm^2

2. a. & b. Drawings will vary. Check the student's drawings.

3. a., b., and c. Answers will vary based on the size of the student's squares. For example:

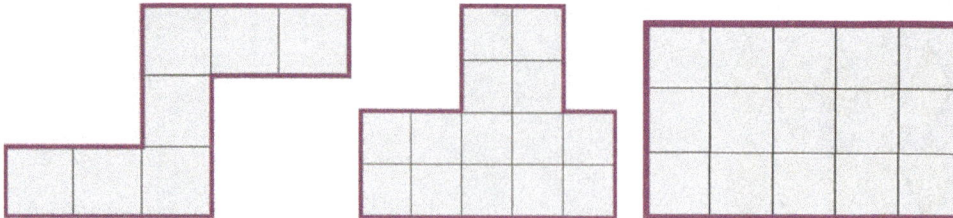

4. Drawings will vary. Check the student's drawings.

5. This other rectangle could have 2 cm and 9 cm sides, or 1 cm and 18 cm sides. If we use fractions of centimetres, it could have for example 12 cm and 1.5 cm sides.

Puzzle corner. a. 20 cm^2 b. Drawings will vary. Check the student's drawings. For example, it could be a right triangle where the two perpendicular sides measure 3 cm and 8 cm, or 4 cm and 6 cm, or 2 cm and 12 cm.

Area of Rectangles 1, pp. 128-130

Teaching Box: The upper rectangle has an area of 30 square units.

A quick way to find the area is to use multiplication. Multiply the number of rows (or the height) by the number of columns (or the length) . In this case: $3 \times 8 = 24$ square units.

1. a. $2 \times 5 = 10$; A = 10 square units   b. $3 \times 3 = 9$; A = 9 square units   c. $3 \times 6 = 18$; A = 18 square units

2. The rectangles can be 1×16, 2×8, or 4×4. For example:

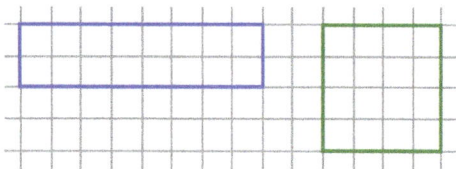

3. You would ask how many columns the rectangle has.

4. See the image on the right.
   a. The rectangle will have 8 columns.
   b. The rectangle will have 4 rows.

5. One column will have 4 squares. $7 \times 4 + 4 = 32$. Its area is 32 square units.

6. a. $8 \times 5 = 40$ square units   b. $7 \times 7 = 49$ square units
   c. $5 \times 9 = 45$ square units

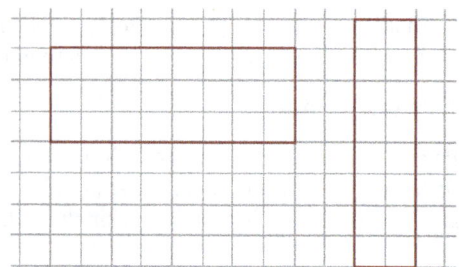

Area of Rectangles 1, cont.

7. a. $5 \times 3 = 15$ square units   b. $2 \times 6 = 12$ square units
   c. $3 \times 5 = 15$ square units   d. $6 \times 1 = 6$ square units

8. The rectangles can be 1×24, 2×12, 3×8, or 4×6.

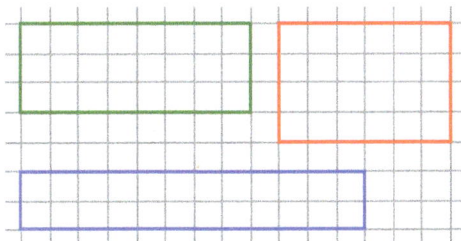

9. The rectangle in (c) can be in the other orientation also.

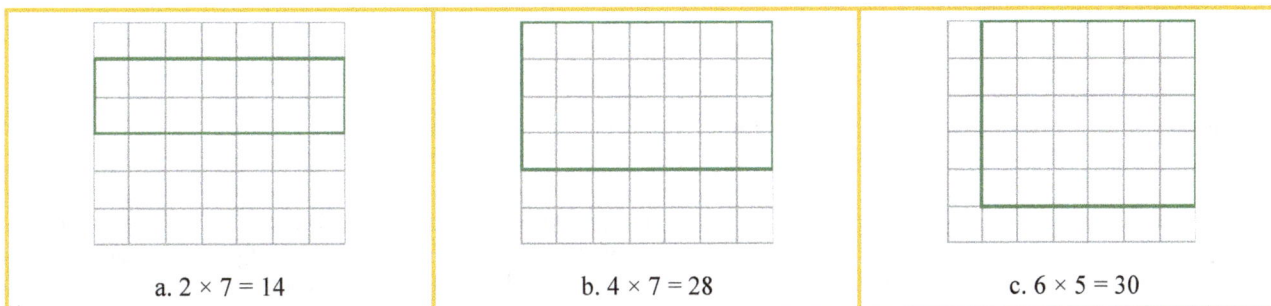

| a. $2 \times 7 = 14$ | b. $4 \times 7 = 28$ | c. $6 \times 5 = 30$ |

Area of Rectangles 2, pp. 131-133

1. a. $5 \text{ cm} \times 1 \text{ cm} = 5 \text{ cm}^2$ b. $6 \text{ cm} \times 3 \text{ cm} = 18 \text{ cm}^2$

2. a. length: __5__ cm   b. length: __4__ cm
   height: __3__ cm   height: __4__ cm
   area: __15__ cm^2 area: __16__ cm^2

3. The area is $9 \text{ cm} \times 2 \text{ cm} = \underline{18 \text{ cm}^2}$

4. a. $8 \text{ cm} \times 2 \text{ cm} = 16 \text{ cm}^2$ b. $4 \text{ cm} \times 3 \text{ cm} = 12 \text{ cm}^2$ c. $10 \text{ cm} \times 5 \text{ cm} = 50 \text{ cm}^2$

5. The other side measures <u>9 cm</u>.

6. a. The 24 cm^2 rectangle can be 2 cm $\times$ 12 cm,   b. The 30 cm^2 rectangle can be 2 cm $\times$ 15 cm,
    3 cm $\times$ 8 cm, or 4 cm $\times$ 6 cm.   3 cm $\times$ 10 cm, or 5 cm $\times$ 6 cm. A 1 cm $\times$ 30 cm
    A 1 cm $\times$ 24 cm rectangle will not fit in this grid.   rectangle will not fit in this grid.

7. Rectangle 1 is $8 \text{ cm} \times 4 \text{ cm} = 32 \text{ cm}^2$. Rectangle 2 is $11 \text{ cm} \times 3 \text{ cm} = 33 \text{ cm}^2$.
   Rectangle 2 is 1 square centimetre larger.

8. No. They are close, but the sticker on the left is $7 \text{ cm} \times 7 \text{ cm} = 49 \text{ cm}^2$, while the one on the right is
   $8 \text{ cm} \times 6 \text{ cm} = 48 \text{ cm}^2$. The sticker on the left is 1 cm^2 larger.

More Units for Measuring Area, pp. 134-135

<u>Page 134</u>

1. a. 12 m^2 b. 56 cm^2 c. 30 km^2

<u>Page 135</u>

2. a. square metres   c. square kilometres
   b. square centimetres   d. square metres

3. It is approximately 15 cm × 5 cm = 75 cm^2.

4.

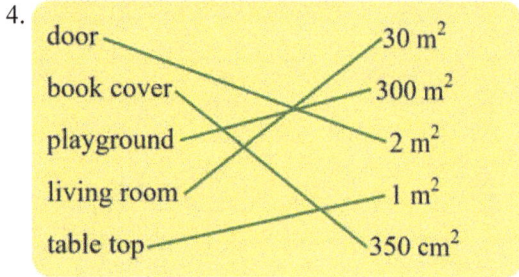

5. Danny's room is 16 m^2. Joe's room is 15 m^2. Danny's room is bigger by one square metre.

Puzzle Corner:

a. Notice that the diagonal line that is one side of the triangle splits the rectangle in half. The rectangle is 7 × 3 = 21 square units. Half of 21 is 10 1/2, so the area of the triangle is 10 1/2 square units.

b. The top point of the triangle is in the middle of the width of the rectangle. So, we can think of this rectangle as being divided into two 4 × 3 rectangles. Again, the diagonal lines are dividing each of those rectangles in half. This makes the shaded area to be exactly half of the area of the entire rectangle. 8 × 3 = 24; 24 ÷ 2 = 12. The area of the triangle is 12 square units.

Area of Compound Shapes 1, pp. 136-137

<u>Page 136</u>

1.

a. 3 × 3 + 3 × 5 = 24 Area = 24 square units	b. 2 × 5 + 3 × 3 = 19 Area = 19 square units
c. 3 × 5 + 2 × 3 = 21 Area = 21 square units	d. 4 × 5 + 2 × 4 = 28 Area = 28 square units

2. 4 × 5 − 4 = 16 square units

<u>Page 137</u>

3. 4 m × 11 m + 4 m × 4 m = 60 m^2

4. 4 × 5 − 3 = 17 square units

5. a. 32 square units   b. 31 square units

6. 10 cm × 15 cm − 6 cm × 4 cm = 150 cm^2 − 24 cm^2 = <u>126 cm^2</u>

Area of Decomposed Rectangles, pp. 138-141

1. a. $3 \times 5 = 15$ square units
   b. $4 \times 5 = 20$ square units
   c. $7 \times 5 = 35$ square units
   d. Each one has 5 as the number of rows.
   e. $5 \times (3 + 4)$ is the area of the entire rectangle.

2. a. See the image on the right.
   b. $3 \times 4 = 12$ square units
   c. $9 \times 4 = 36$ square units
   d. $12 \times 4 = 48$ square units
   e. $4 \times (3 + 9)$ is the area of the entire rectangle.

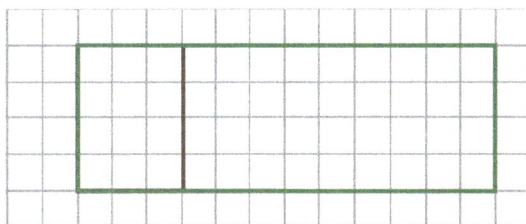

Page 139

3.

a.

$3 \times (3 + 5)$ = 3×3 + 3×5

area of the whole rectangle   area of the first part   area of the second part

b.

$3 \times (4 + 2)$ = 3×4 + 3×2

area of the whole rectangle   area of the first part   area of the second part

c.

$2 \times (3 + 3)$ = 2×3 + 2×3

area of the whole rectangle   area of the first part   area of the second part

4.

a. $3 \times (2 + 4) = 18$	b. $5 \times (1 + 4) = 20$

Area of Decomposed Rectangles, cont.

5.

a.

$4 \times (3 + 1)$ = __4__ × __3__ + __4__ × __1__

area of the   area of the   area of the
whole rectangle   first part   second part

b.

__3__ × (__2__ + __1__) = 3×2 + 3×1

c.

__2__ × (__5__ + __2__) = 2×5 + 2×2

6. a. $4 \times (4 + 2) = 4 \times 4 + 4 \times 2$

b. $6 \times (3 + 2) = 6 \times 3 + 6 \times 2$

Page 141

7. See the image on the right.
Example explanation: The whole 6×11 rectangle has the same area as the 6×2 and 6×9 rectangles put together: 66 square units. Also, the side lengths are the same for the whole rectangle as with the two rectangles put together.

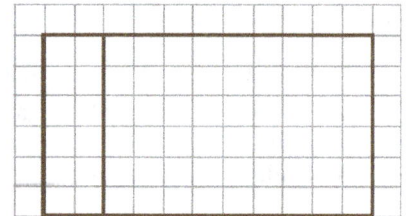

8. Expressions (i) and (iv) match the model.

9.

| a. $5 \times (7 + 8)$ | b. $7 \times 1 + 7 \times 6$ |

Puzzle Corner:

a. One way to reason it out is this.
The equation simplifies to $5 \times (2 + b) = 50$, from which we can see that $2 + b$ must equal 10. So, $b = 8$.

b. In this case, we need to first find out what number times 8 gives us 104. By guess and check, students can find out it is 13. So, we have: $8 \times 13 = 8 \times (s + 7)$. Then we see that $13 = s + 7$ from which $s = 6$.

Multiplying by Multiples of Ten, pp. 142-143

Page 142

1. Break the rectangle that is 7 units by 20 units into two parts, each part being 7 units by 10 units.
The area of each part is 70 square units, thus the area of the entire rectangle is twice that, or 140 square units.

2. $5 \times 30 = 150$. The image suggests we can solve this by breaking 5×30 into 5×10 and 5×10 and 5×10, and adding those.

Multiplying by Multiples of Ten, cont.

3.

a. 7×90 $= \underline{7} \times \underline{9} \times 10$ $= \underline{63} \times 10 = 630$	b. 4×80 $= 4 \times 8 \times 10$ $= 32 \times 10 = 320$	c. 6×40 $= 6 \times 4 \times 10$ $= 24 \times 10 = 240$
d. 9×90 $= 9 \times 9 \times 10$ $= 81 \times 10 = 810$	e. 30×12 $= 10 \times 3 \times 12$ $= 10 \times 36 = 360$	f. 80×3 $= 10 \times 8 \times 3$ $= 10 \times 24 = 240$

4. a. 490   b. 480   c. 440   d. 720
     120   600   350   240

5. The area is $7 \times 80 = 560$ square units.

6. $7 \times 10 = 70$ square units

7. The total area: $8 \times 30 = 240$ square units. Area of each part: $8 \times 10 = 80$ square units.

8. The rectangle is divided into thirds. Each third has the area of $7 \times 40 = 280$ square units. The total area is then $280 + 280 + 280 = 840$ square units.

Puzzle corner. a. 600   b. 1600   c. 9600   d. 2000   e. 6000

Area of Compound Shapes 2, pp. 144-145

1. There are several ways to break this shape into three rectangles. For example:
     $8 \times 9 + 6 \times 6 + 3 \times 4 = 120$ square units
   or $8 \times 5 + 17 \times 4 + 6 \times 2 = 120$ square units

2. The student may use a different way of breaking the shape into rectangles. One way is shown in the image on the right. The top rectangle is 6 m × 4 m. The middle rectangle is $(3 \text{ m} + 6 \text{ m}) \times (10 \text{ m} - 4 \text{ m}) = 9 \text{ m} \times 6 \text{ m}$. The bottom rectangle is 6 m × 3 m. So the total area of the floor plan is:
   $6 \text{ m} \times 4 \text{ m} + 9 \text{ m} \times 6 \text{ m} + 6 \text{ m} \times 3 \text{ m} = \underline{96 \text{ m}^2}$.

3. $(2 \text{ m} + 4 \text{ m}) \times 2 \text{ m} = 12 \text{ m}^2$

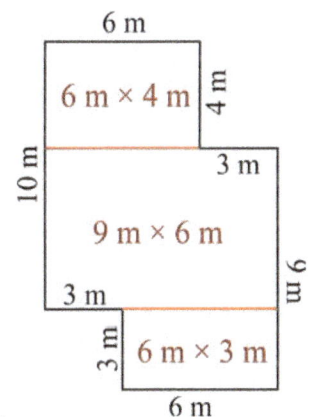

4. a. 39 square units   b. 29 square units   c. 39 square units   d. 32 square units

5. The student may use different ways to break the shapes down, so their number sentences may vary from the ones below. They should still arrive at the same total.
   a. $3 \times 3 + 4 \times 3 + 6 \times 3 = 39$ square units   b. $8 \times 5 - 3 \times 1 - 2 \times 2 - 2 \times 2 = 29$ square units
   c. $8 \times 6 - 3 \times 3 = 39$ square units   d. $4 \times 3 + 4 \times 6 - 1 \times 4 = 32$ square units

6. $30 \text{ m} \times 12 \text{ m} - 10 \text{ m} \times 6 \text{ m} = 300 \text{ m}^2$

Puzzle Corner:

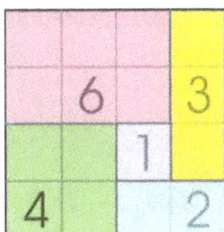

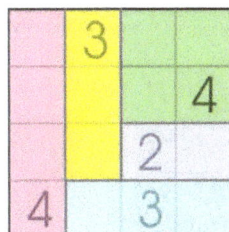

Perimeter, pp. 146-148

Page 146

Teaching box: The perimeter of the blue rectangle is 12 units. The perimeter of the shape in Example 1 is 16 units.

1. a. 14 units   b. 12 units   c. 12 units
   d. 12 units   e. 18 units   f. 24 units

Page 147

Example 2. The perimeter of the triangle is 27 units.

2. a. 49 cm   b. 24 units
   c. 12 cm   d. 86 cm

3. 7 cm + 7 cm + 9 cm + 9 cm = 32 cm

Page 148

4. a. 16 cm   b. 12 cm
   c. 13 cm   d. 26 cm

5. Please check the student's measurements. The measurements below may not match yours or the student's, if you printed this page yourself and used "shrink to fit" or a similar printer setting.

Side AB   __58__ mm
Side BC   __49__ mm
Side CD   __42__ mm
Side DA   __45__ mm
Perimeter   __194__ mm

Puzzle Corner:
A regular octagon has eight equal sides.
8 × 61 cm = 488 cm; Each side is 61 cm.

More About Perimeter, pp. 149-150

Page 149

1. a. $6 + x + 6 + x = 20$ or $6 + x = 10$. Solution: $x = 4$ cm.
  b. $15 + y + 15 + y = 44$ or $15 + y = 22$. Solution: $y = 7$ cm.
  c. $9 + 8 + 12 + 5 + s = 42$. Solution: $s = 8$ m.
  d. $s + s + s + s = 12$ or $4 \times s = 12$. Solution: $s = 3$ cm.

Page 150

2. Since we know that these shapes can be broken into rectangles, we can determine that the total lengths of opposite sides will be equal.

a. For both unknown sides, the opposite sides are 6 cm, and the known length from the same side is 3 cm. So each unknown side is 6 − 3 = 3 cm.

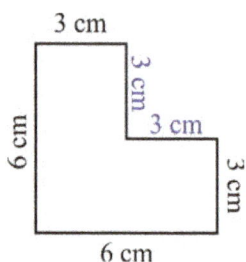

The total perimeter is:
6 cm + 6 cm + 3 cm + 3 cm + 3 cm + 3 cm = __24 cm__.

b. There are three side lengths not given. The bottom can be determined by adding the lengths of the parallel lines on top: 8 + 5 + 4 = 17. The other two are the same as the other two vertical lines: 8 m & 4 m.

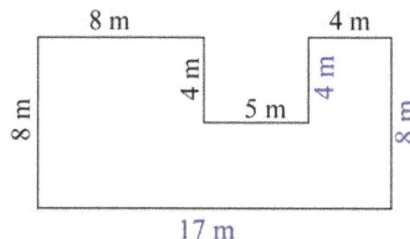

8 + 8 + 4 + 5 + 4 + 4 + 8 + 17 = 58.
The perimeter is __58 m__.

3. Student drawings will vary. For example:

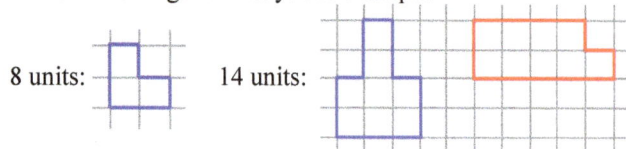

8 units: 14 units:

4. 6 × 11 = 66; She needs 66 centimetres of frame border.

5. a. Counting the units in the picture, the perimeter is 18 units.
  b. Since each unit is 10 metres, we get 18 × 10 metres = 180 metres.

Area and Perimeter Problems, pp. 151-152

Page 151

1. a. perimeter 14 m; area 10 m^2
   b. perimeter 24 km; area 36 km^2
   c. perimeter 12 cm; area 8 cm^2
   d. perimeter 12 cm; area 9 cm^2

2. You can divide the shape into four 4 m by 4 m squares, each having the area of 16 m^2. The area is then
   16 m^2 + 16 m^2 + 16 m^2 + 16 m^2 = 64 m^2.
   The perimeter is 40 m.

Page 152

3. For the area, divide the shape into two rectangles. That can be done in two ways. You could get
   11 cm × 4 cm + 4 cm × 8 cm = 76 cm^2.
   or 4 cm × 12 cm + 7 cm × 4 cm = 76 cm^2.

   The perimeter is
   4 cm + 8 cm + 7 cm + 4 cm + 11 cm + 12 cm = 46 cm.

Page 152

4. a. 5 m × 4 m = 20 m^2 and 10 m × 4 m = 40 m^2
   b. 60 m^2
   c. 38 m

5. Area of each little part is 6 m × 10 m = 60 m^2.
   The total area is 6 m × 60 m = 360 m^2.

Puzzle corner.
   a. a 13 × 3 rectangle.

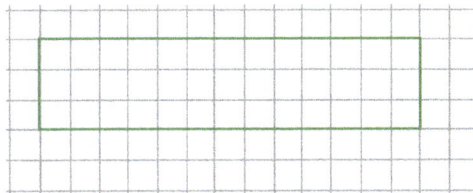

   b. a 14 × 4 rectangle.

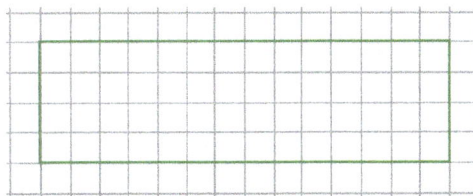

Same Area, Different Perimeter, pp. 153-154

Page 153

1. Each rectangle listed in the table can also have the Side 1 and Side 2 switched. (e.g. instead of 2 units and 12 units, it would be 12 units and 2 units).

Rectangle	Side 1	Side 2	Area	Perimeter
Rectangle 1	2 units	12 units	24 square units	28 units
Rectangle 2	3 units	8 units	24 square units	22 units
Rectangle 3	4 units	6 units	24 square units	20 units
Rectangle 4	1 unit	24 units	24 square units	50 units

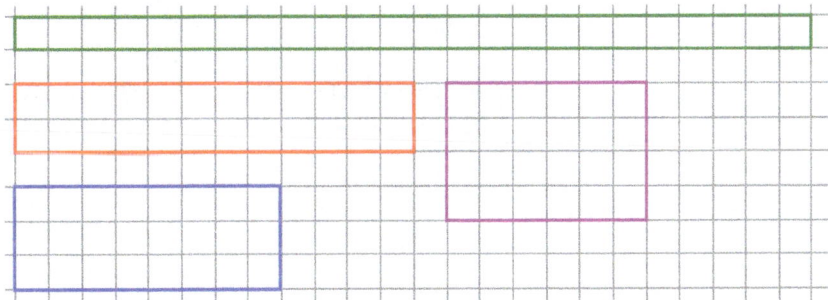

2. a. The area of the coop should be 12 m^2.
   b. A rectangle of 1 m × 12 m is 12 m^2, but it is very narrow. Other possible sizes are 3 m × 4 m and 2 m × 6 m.

Same Area, Different Perimeter, cont.

Page 154

3. The largest possible perimeter is 62 units.

Rectangle	Side 1	Side 2	Area	Perimeter
1	1 units	30 units	30 square units	62 units
2	2 units	15 units	30 square units	34 units
3	3 units	10 units	30 square units	26 units
4	5 unit	6 units	30 square units	22 units

4. a. 3 m × 12 m, 4 m × 9 m, and 6 m × 6 m are the most likely possibilities.
   2 m × 18 m and 1 m × 36 m would be very narrow rooms.
   b. 6 m × 6 m gives the smallest perimeter: 24 m.

5. a. There are numerous possibilities, but we will start with a minimum width of 6 m:
   6 m × 14 m, 6 m × 15 m, 6 m × 16 m, 7 m × 12 m, 7 m × 13 m, 7 m × 14 m,
   8 m × 11 m, 8 m × 12 m, 9 m × 9 m, 9 m × 10 m, 10 m × 10 m
   b. 9 m × 9 m and 8 m × 10 m both have the smallest possible perimeter: 36 m.

Same Perimeter, Different Area, pp. 155-156

Page 155

1. Each rectangle listed in the table can also have the Side 1 and Side 2 switched.

	Side 1	Side 2	Perimeter	Area
Rectangle 1	2 units	8 units	20 units	16 square units
Rectangle 2	3 units	7 units	20 units	21 square units
Rectangle 3	4 units	6 units	20 units	24 square units
Rectangle 4	5 units	5 units	20 units	25 square units
Rectangle 5	1 unit	9 units	20 units	9 square units

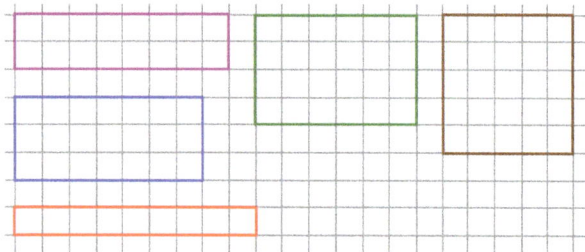

2. The largest area is 25 square units. The smallest is 9 square units.

3. Students will likely have to investigate this and try different side lengths, until they find out that a square with a given perimeter has the largest area. In this case, the largest area it can have is 64 square units (8 × 8).

Page 156

4. Drawings may vary. Rectangles could be 3 m × 5 m or 4 m × 4 m.

5. a. The other side is 1 cm.
   b. The area is 6 cm^2.
   c. It should be 2 cm × 3 cm.

6. The rectangle should be 4 units by 9 units.

7. 2 × (5 + 2) = 2 × 5 + 2 × 2   See the image on the right:

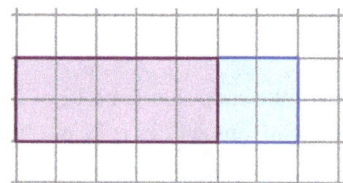

Puzzle Corner: Notice that each square unit is 2 cm × 2 cm. The perimeter is 44 cm. The area is 72 cm^2.

1. 11 cm

2. The image on the right shows all the side lengths.
   600 + 1200 + 600 + 480 + 240 + 240 + 240 + 480 = 4080;
   The perimeter is <u>4080 cm</u>.

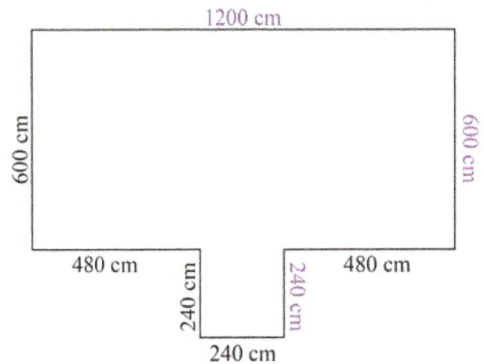

3. 6 m

4. a. 9 m × 20 m = 180 m^2 and 9 m × 30 m = 270 m^2
   b. 450 m^2
   c. 118 m

5. a. 20 m × 3 m = 60 m^2
   b. 180 m^2
   c. 58 m

6.

a. 3 × (5 + 2) = 3 × 5 + 3 × 2	
b. 3 × (2 + 1) = 3 × 2 + 3 × 1	

Puzzle corner:
   a. Since the long side measures 50 m, the height must be <u>9 m</u>, because 9 m × 50 m = 450 m^2.
   b. Either add the numbers in brackets, and then multiply by 20, or
      multiply each number by 20 and then add those totals together: 280.

Mixed Revision Chapter 10, pp. 159-162

1. a. She has 40 minutes left.
   b. It took him 28 minutes.
   c. The class ends in 13 minutes.

2. Area: 12 cm^2
   Perimeter: 16 cm

3. 6 × 7 + 3 = 45; He had 45 toy cars.

4.

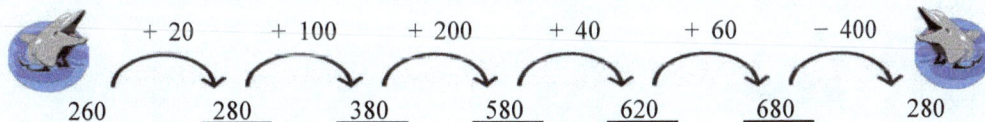

Page 160

5. a. _____
   b. _____
   c. _____

6. Answers will vary. Check the student's measurements and the line plot.

7. a. The cheaper refrigerator is $477.
   b. His total cost was $954.
   c. His change was $46.

Page 161

8. a. 70 kg   b. 120 g   c. 22 kg   d. 90 g

9. Student graphs will vary since the student may use a different scale for the dollar amounts. Check the student's graph.

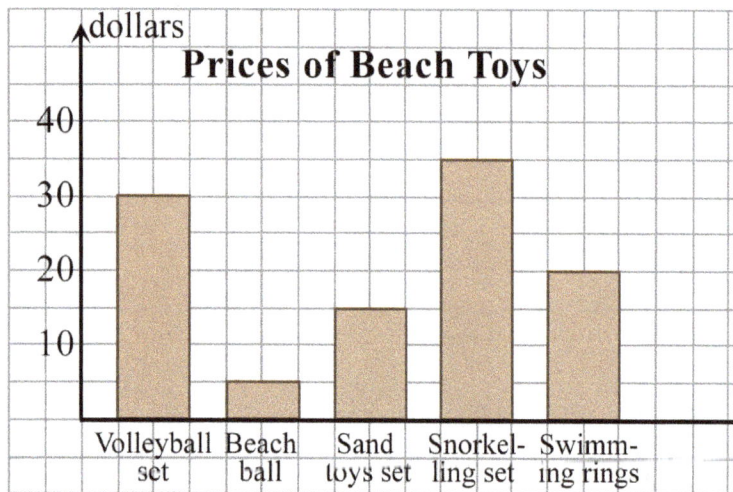

10. a. It costs $15 more.
    b. They cost $40 together.
    c. The total is $20.

Page 162

11. The time now: a. 2:48   b. 5:23   c. 1:57
    15 min later: 3:03   5:38   2:12

12. a. 5637   b. 6121
    c. 9696   d. 4010

13. a. 0, 0   b. 12, 8   c. 0, 0   d. 150, 420

Puzzle Corner: There are many possible answers. These are just some examples.

80	÷	8	=	10
÷		÷		
10	÷	2	=	5
=		=		
8		4		

54	÷	9	=	6
÷		÷		
6	÷	3	=	2
=		=		
9		3		

Revision Chapter 10, pp. 163-165

<u>Page 163</u>

1. a. A, B, F, H, J   b. C, E, I, K, L

2. Answers will vary. Check the student's answers. The image on the right shows one example.

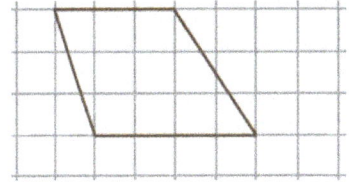

3. a. Shape 5, a rectangle   d. Shape 1, a regular hexagon
   b. Shape 3, a rhombus   e. Shape 4
   c. Shape 2   f. Shape 6

4. 48 ft^2

<u>Page 164</u>

5. Rectangles can be 1 × 20, P = 42; 2 × 10, P = 24; or 4 × 5, P = 18.

6. a. A = 3 × 2 + 3 × 4 = 18 square units   b. A = 2 × 2 + 3 × 4 = 16 square units

7. There are different ways to break the shape into rectangles, but the total area should be the same. Starting from the lower left: 4 cm × 2 cm = 8 cm^2; 2 cm × 5 cm = 10 cm^2; 4 cm × 2 cm = 8 cm^2; 3 cm × 5 cm = 15 cm^2. Total area: 41 cm^2

8. a. 490   b. 480   c. 280

9. Area of each part: 9 × 10 = 90 square units. Total area 9 × 40 = 360 square units.

<u>Page 165</u>

10.

a.	
3 × (5 + 1) = 3 × 5 + 3 × 1	

b.	
4 × (2 + 3) = 4 × 2 + 4 × 3	

11. Area 35 cm^2
    perimeter 24 cm

12. 12 + 12 + w + w = 82
    Solution: w = 29 cm

13. a. Area: 3 × 4 = 12 square units   b. Area: 11 square units
    Perimeter: 3 + 3 + 4 + 4 = 14 units   Perimeter: 24 units

Chapter 11: Fractions

Understanding Fractions 1, pp. 170-173

Page 171

1. a. 1/4   b. Not equally divided   c. Not equally divided. (However, some students might notice that if you draw one more line through the square, you can divide it into eight equal parts. The shaded part is then 2/8 or 1/4 of the whole.)
   d. Not equally divided   e. 1/2

2. a. No   b. 1/6   c. No   d. No   e. 1/4

3. Answers will vary because there are often several ways to divide a shape into equal parts. For example:

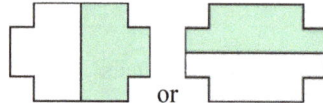

a. four equal parts	b. three equal parts	c. eight equal parts	d. six equal parts
$\frac{1}{4}$	$\frac{1}{3}$	$\frac{1}{8}$	$\frac{1}{6}$

e. four equal parts	f. two equal parts	g. eight equal parts	h. three equal parts
$\frac{1}{4}$	$\frac{1}{2}$	$\frac{1}{8}$	$\frac{1}{3}$

Page 172

4. Student activity. See the illustrations in the worktext.

5. The shapes *c* and *e* have six equal parts, with one part shaded.

6.

Student 1: $\frac{1}{8}$	Incorrect. The answer given is the ratio of shaded parts to non-shaded parts. The correct fraction is 1/9.	Student 2: $\frac{1}{3}$	Incorrect. The parts are not all the same size. The shaded part is smaller than the non-shaded parts, so it is less than 1/3.
Student 3: $\frac{1}{5}$	Correct.	Student 4: $\frac{1}{2}$	Incorrect. The answer given is the ratio of shaded parts to non-shaded parts. The correct fraction is 1/3.
Student 5: $\frac{1}{9}$	Incorrect. The triangle parts on the ends of the shape are half the size of the rest of the triangle parts, so the shaded part is less than 1/9.	Student 6: $\frac{1}{2}$	Incorrect. The parts must be equal shape and size.

Understanding Fractions 1, cont.

7. a. Not equal parts   b. 1/8   c. Not equal parts

8. Answers will vary because there often are several ways to divide a shape into equal parts. For example:

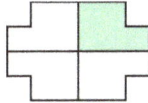

a. six equal parts	b. four equal parts	c. eight equal parts	d. two equal parts
$\dfrac{1}{6}$	$\dfrac{1}{4}$	$\dfrac{1}{8}$	$\dfrac{1}{2}$

e. eight equal parts	f. four equal parts	g. six equal parts	h. two equal parts
$\dfrac{1}{8}$	$\dfrac{1}{4}$	$\dfrac{1}{6}$	$\dfrac{1}{2}$

For 8. b., on the right is another way to divide it into four equal parts. This is based on the fact the shape itself can be divided into 20 square units.

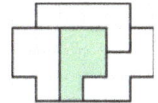

Puzzle Corner:
Yes. The large rectangle is divided into four equal rectangles, and each of those are divided equally in half. So there are eight equal parts, and the one that is shaded is 1/8 part of the shape.

Understanding Fractions 2, pp. 174-177

1. The colouring will vary. Any parts of the whole can be coloured. For example:

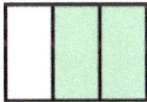

a. 2 thirds	b. 4 sixths	c. 3 fifths	d. $\dfrac{3}{8}$	e. $\dfrac{2}{2}$	f. $\dfrac{9}{10}$

2. a. 1/3; one-third   b. 2/5; two-fifths   c. 3/4; three-fourths   d. 2/8; two-eighths
   e. 5/5; five-fifths   f. 3/6; three-sixths   g. 5/10; five-tenths   h. 7/8; seven-eighths

3.

a. $\dfrac{2}{3}$	b. $\dfrac{2}{5}$	c. $\dfrac{3}{6}$	d. $\dfrac{6}{8}$
e. $\dfrac{4}{5}$	f. $\dfrac{3}{8}$	g. $\dfrac{3}{3}$	h. $\dfrac{7}{8}$

113

Understanding Fractions 2, cont.

4.

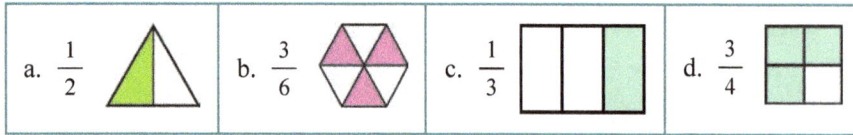

5.

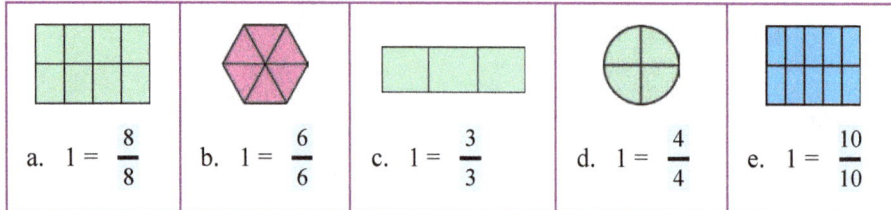

Teaching box: You would need 8 fourths to fill two whole tables.

6.

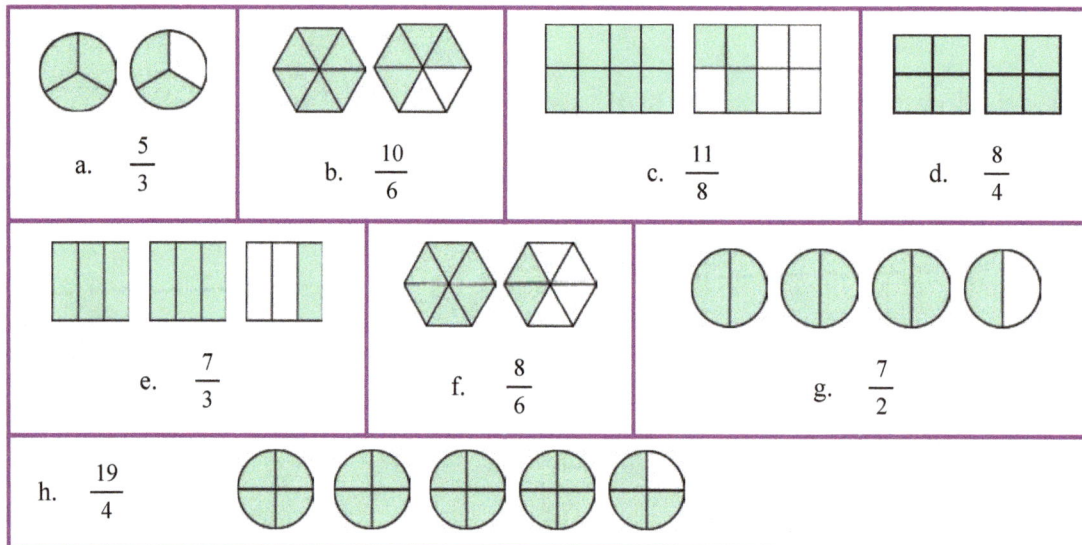

7. a. 1 half, 2 halves = 1 whole,   3 halves, 4 halves = 2 wholes,
    5 halves, 6 halves = 3 wholes,   7 halves, 8 halves = 4 wholes.

  b. 1 third, 2 thirds, 3 thirds = 1 whole,   4 thirds, 5 thirds, 6 thirds = 2 wholes,
    7 thirds, 8 thirds, 9 thirds = 3 wholes, 10 thirds, 11 thirds, 12 thirds = 4 wholes.

  c. 1 fourth, 2 fourths, 3 fourths, 4 fourths = 1 whole,
    5 fourths, 6 fourths, 7 fourths, 8 fourths = 2 wholes,
    9 fourths, 10 fourths, 11 fourths, 12 fourths = 3 wholes,
    13 fourths, 14 fourths, 15 fourths, 16 fourths = 4 wholes.

  d. 1 fifth, 2 fifths, 3 fifths, 4 fifths, 5 fifths = 1 whole,
    6 fifths, 7 fifths, 8 fifths, 9 fifths, 10 fifths = 2 wholes,
    11 fifths, 12 fifths, 13 fifths, 14 fifths, 15 fifths = 3 wholes,
    16 fifths, 17 fifths, 18 fifths, 19 fifths, 20 fifths = 4 wholes.

8. a. 7/4; seven-fourths   b. 8/6; eight-sixths   c. 8/3; eight-thirds   d. 4/2; four-halves
   e. 5/3; five-thirds   f. 10/4; ten-fourths   g. 24/8; twenty four-eighths

Understanding Fractions 2, cont.

9. There are several ways to draw the fractional parts together to make one whole. For example:

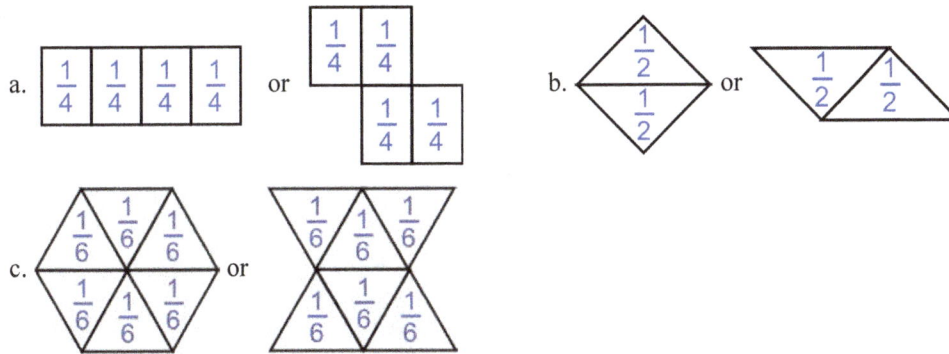

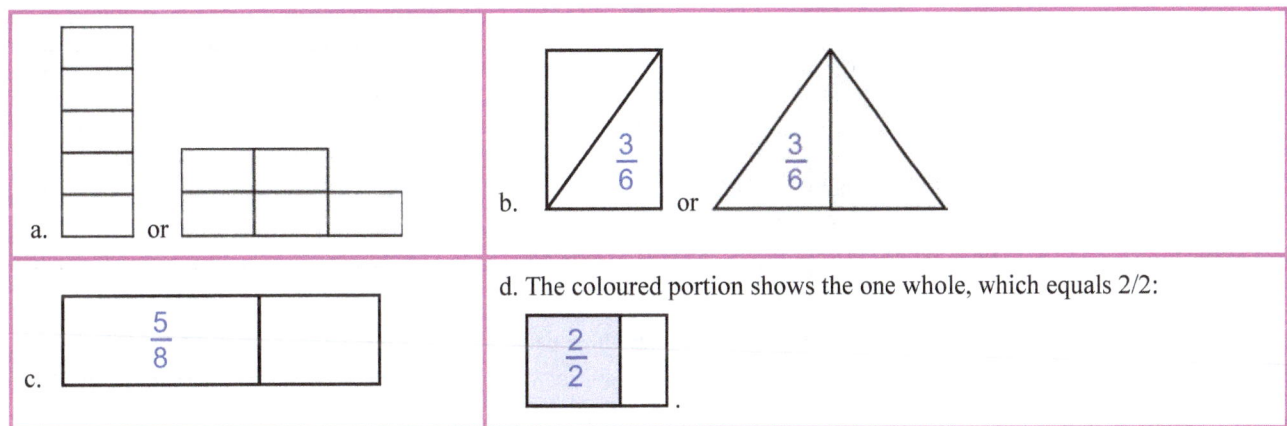

Puzzle Corner: answers will vary. For example:

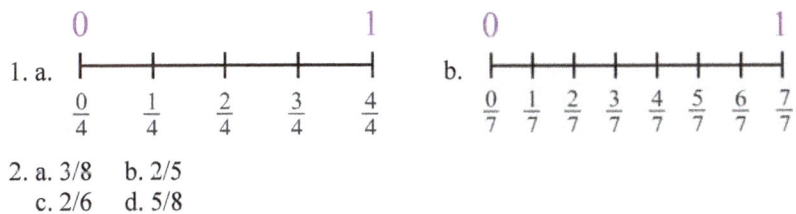

Fractions on the Number Line 1, pp. 178-180

2. a. 3/8   b. 2/5
   c. 2/6   d. 5/8

3. a. The top number line is divided into sixths.

5. a. 1/3   b. 2/9
   c. 6/10   d. 4/6

115

Fractions on a Number Line, cont.

6. a. The dot marks 3/8.   b. He most likely counted the zero mark as the first fractional part.

7. a. 1/4   b. 1/3

c. 3/4   d. 1/6

e. 5/6   f. 3/8

g. 4/5   h. 5/8

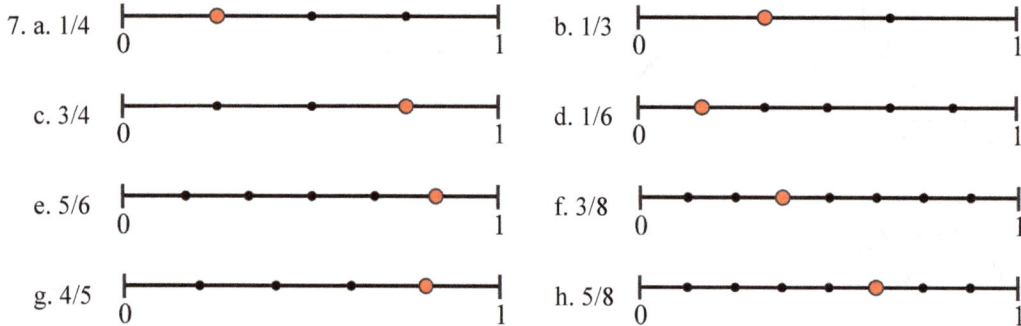

8. $\dfrac{5}{5}$ is the biggest fraction. $\dfrac{5}{10}$ is the smallest fraction. See the number lines on the right.

$\dfrac{5}{6}$

$\dfrac{5}{8}$

9. The correct way to show a fraction on a number line is to mark the fraction on the mark, as it is on the top line. This is because in a number line illustration, the point for a number shows the distance from zero to that point.

$\dfrac{5}{5}$

If we had a rectangle (not a number line) that was divided into three parts, then labelling the section of one-third below it (like in the bottom illustration) would be correct.

$\dfrac{5}{10}$

Fractions on a Number Line 2, pp. 181-184

1. a.

| 0 | 1 | 2 | 3 |

$\dfrac{3}{6}$ $\dfrac{7}{6}$ $\dfrac{11}{6}$ $\dfrac{13}{6}$ $\dfrac{18}{6}$

18/6 is the whole number 3.

b.

| 0 | 1 | 2 | 3 |

$\dfrac{6}{5}$ $\dfrac{9}{5}$ $\dfrac{11}{5}$ $\dfrac{13}{5}$ $\dfrac{15}{5}$

15/5 is the whole number 3.

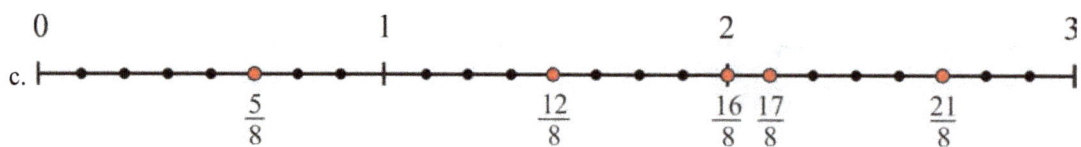

c.

| 0 | 1 | 2 | 3 |

$\dfrac{5}{8}$ $\dfrac{12}{8}$ $\dfrac{16}{8}$ $\dfrac{17}{8}$ $\dfrac{21}{8}$

16/8 is the whole number 2.

d. 2 = 12/6   e. 3 = 24/8

116

Fractions on a Number Line 2, cont.

2.
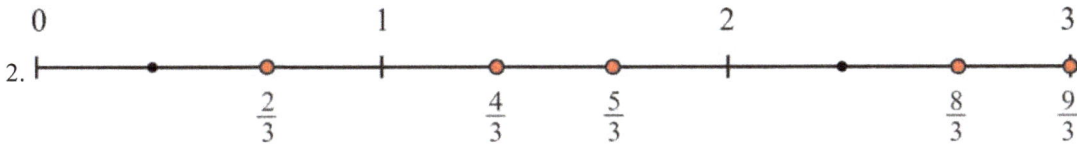

3. 4 = 12/3

4. a. 1 = 6/6   b. 2 = 16/8   c. 3 = 12/4
   d. 2 = 12/6   e. 3 = 15/5   f. 3 = 9/3   g. 2 = 16/8   h. 3 = 15/5

5. The number of fractional parts in one whole is the denominator (bottom number), and you can multiply the number of wholes by the denominator to find the top number (numerator).
   8/8 is one whole, in eighths, so 5 wholes would be $(5 \times 8)/8 = 40/8$.
   3/3 is one whole, in thirds, so 5 wholes would be $(5 \times 3)/3 = 15/3$.

6.
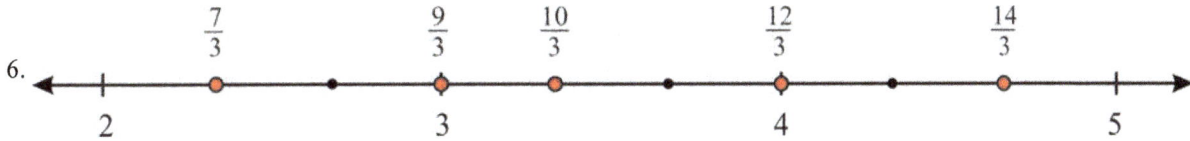

7.
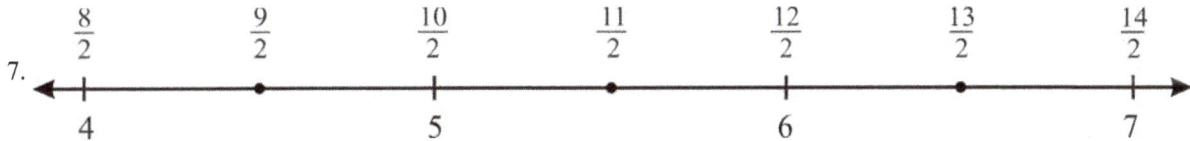

8.
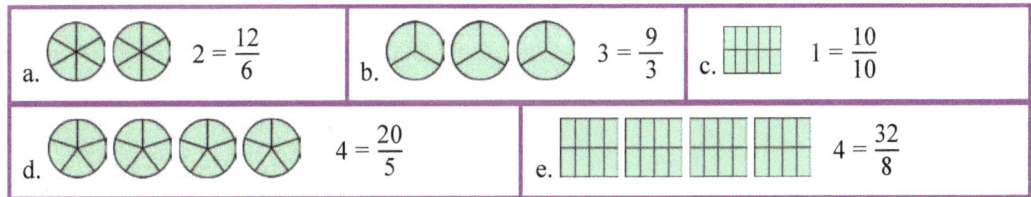

9.
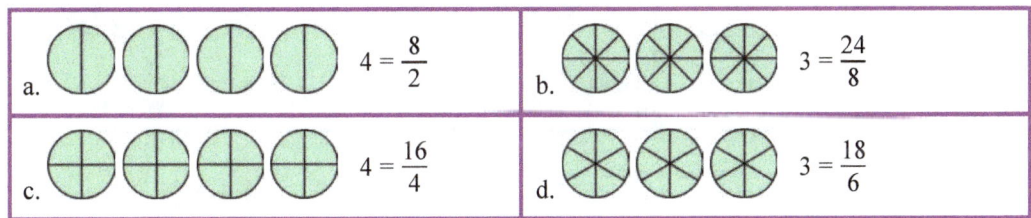

10. Any fraction where the numerator divided by the denominator equals two, will work.
    Examples: 4/2, 10/5, 20/10

11. a. 1   b. 3   c. 4   d. 10
    e. 5   f. 1   g. 4   h. 6

12.
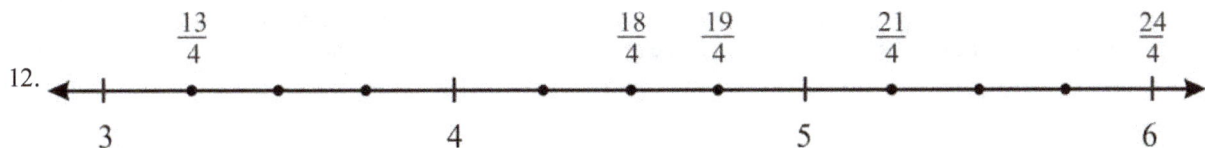

Fractions on a Number Line 2, cont.

Page 184

13.

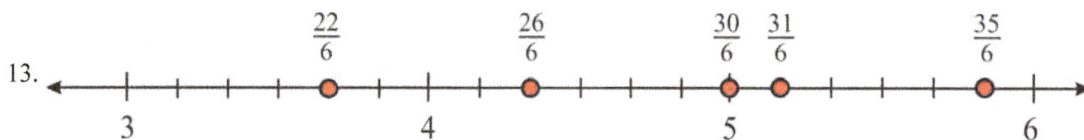

14.

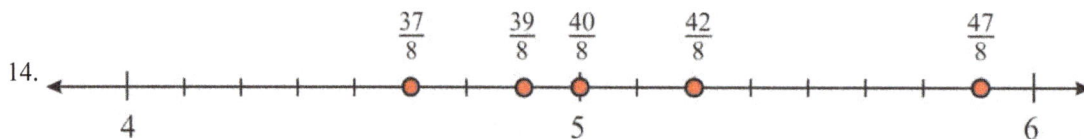

15. Neither is correct. The whole number 8 equals 64/8 and 9 equals 72/8. So, the fraction 69/8 is actually between 8 and 9.

16. 36/3 is 36 ÷ 3, which equals 12 and is the greatest number in this set.

Puzzle Corner: First, figure out where 1 is, by dividing the interval from 0 to 3-4 into three parts and thus finding the points for 1/4 and 1/2. After that, you can mark the points by fourths.

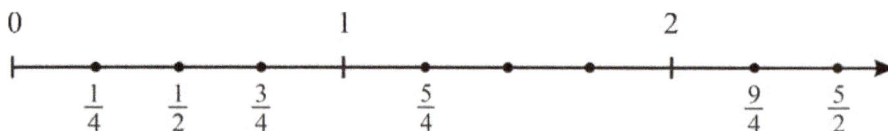

Equivalent Fractions 1, pp. 185-186

Page 185

1. a. $\frac{4}{6} = \frac{2}{3}$ b. $\frac{2}{3} = \frac{6}{9}$ c. $\frac{3}{6} = \frac{1}{2}$

   d. $\frac{1}{3} = \frac{3}{9}$ e. $\frac{6}{8} = \frac{3}{4}$

2.

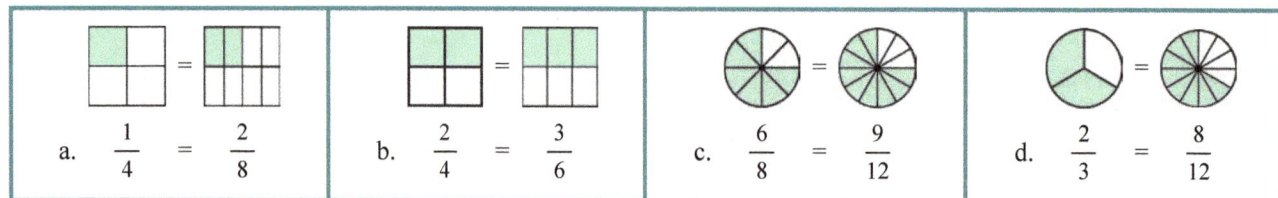

a. $\frac{1}{4} = \frac{2}{8}$ b. $\frac{2}{4} = \frac{3}{6}$ c. $\frac{6}{8} = \frac{9}{12}$ d. $\frac{2}{3} = \frac{8}{12}$

Page 186

3. Illustrations may vary. For example:

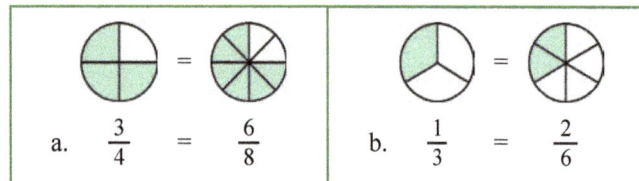

a. $\frac{3}{4} = \frac{6}{8}$ b. $\frac{1}{3} = \frac{2}{6}$

4. Fractions and illustrations may vary. Students may use pie models, rectangular models, number lines, squares, or other shapes to show this. The denominator should be twice as large as the numerator. Examples:

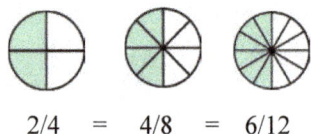

2/4 = 4/8 = 6/12

Equivalent Fractions, cont.

5. *a*, *b*, and *c* all show fractions equivalent to 3/4.

6. Yes. Both 3/3 and 4/4 are equal to 1, so they are equivalent fractions.

7.

a. $\dfrac{3}{4} = \dfrac{6}{8}$	b. $\dfrac{3}{5} = \dfrac{6}{10}$	c. $\dfrac{1}{3} = \dfrac{2}{6}$
d. $\dfrac{1}{4} = \dfrac{2}{8}$	e. $\dfrac{1}{3} = \dfrac{3}{9}$	f. $\dfrac{3}{4} = \dfrac{9}{12}$

Equivalent Fractions 2, pp. 187-188

1.

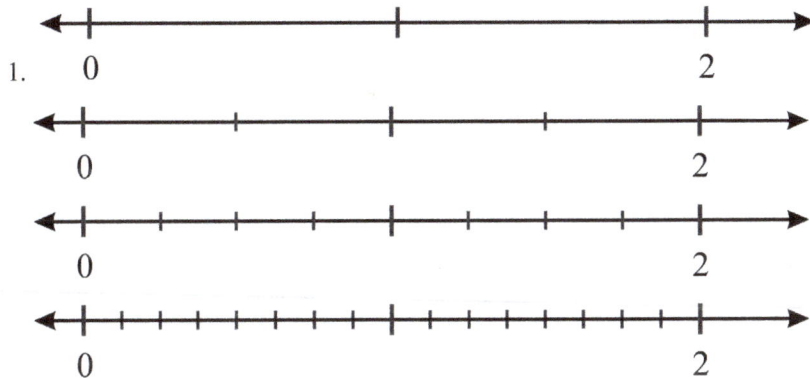

2. Fractions may vary. The numerator divided by the denominator should equal two. Students may use pie models, rectangular models, number lines, squares, or other shapes to show this. If the student has difficulty, point out that the number lines from question #1 can help. Examples: 2/1, 4/2, 8/4, 6/3, 10/5, 14/7, 16/8, etc.

3. a. 1/1 = 2/2 = 4/4 = 8/8   b. 1/4 = 2/8   c. 3/2 = 6/4 = 12/8

4. Fractions may vary for *b*.

a.	$\dfrac{3}{4}$	b.	$\dfrac{2}{3}$
	$\dfrac{6}{8}$		$\dfrac{4}{6}$
	$\dfrac{9}{12}$		$\dfrac{6}{9}$
	$\dfrac{12}{16}$		$\dfrac{8}{12}$

5. a. 6/2, 9/3, 3/1

   b. They can be determined to all be equal to 3, because each one, the numerator divided by the denominator equals 3.

6. Amanda ate half of her pizza. For Joe to eat half of his pizza, he would eat half of the 8 pieces. He ate 4 pieces.

Equivalent Fractions 2, cont.

7. The matching fractions are:
   1/4: Top row: 2nd, 4th, bottom row: 5th.
   1/3: Top row: 3rd, bottom row: 3rd.
   3/4: Top row: 5th, bottom row: 4th.
   2/3: Bottom row: 1st, 2nd.

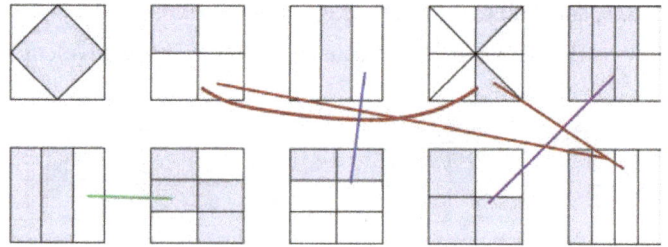

Equivalent Fractions 3, pp. 189-190

1. 1/3 is equivalent to it.

2. The way the student corrects the false equations will vary.

   a. Not correct. Corrected version: 3/3 = 1 or 9/3 = 3 or 3/1 = 3.
   b. Not correct. Corrected version: 3/1 = 3 or 1/1 = 1 or 3/3 = 1.
   c. Correct.
   d. Correct.
   e. Not correct. Corrected version: 3 = 9/3 or 3 = 12/4 or 4 = 12/3.

3. Illustrations will vary; check the student's work. For example:

4. One easy way to show Liam he is wrong is by drawing pictures of the two fractions:

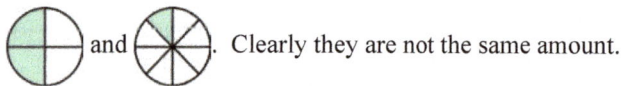 Clearly they are not the same amount.

An equivalent fraction is not formed by multiplying the numerator and the denominator. However, perhaps Liam is thinking of the equivalence of 1/4 and 2/8, and how, if you multiply both 1 and 4 by two, you get 2 and 8. That is a true process: you *can* multiply both the numerator and the denominator by some same number and that way get an equivalent fraction.

A fraction is the same as a division, so 2/4 can't be calculated as 2 × 4. 2/4 is the same as 2 divided by 4, and 1/8 is 1 divided by 8, which are not equivalent.

5. a.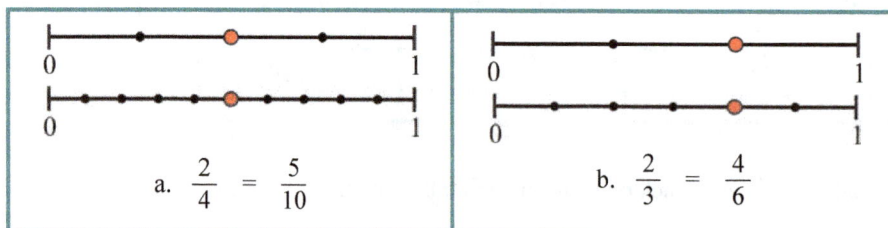
   b. 1/2 = 3/6

6.

$0 \quad\quad\quad 1$ $0 \quad\quad\quad 1$ a. $\dfrac{2}{4} = \dfrac{5}{10}$	$0 \quad\quad\quad 1$ $0 \quad\quad\quad 1$ b. $\dfrac{2}{3} = \dfrac{4}{6}$

Equivalent Fractions 3, cont.

7. a. 8/2

  b. Answers may vary. If the bars are cut into thirds, the fraction will be 12/3. Or, like in the image below, if they are cut into fourths, the fraction will be 16/4.

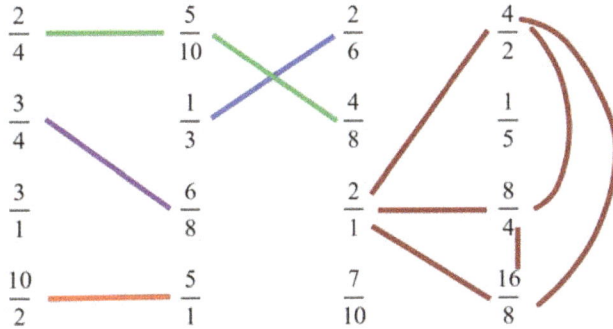

8. 2/4 = 5/10 = 4/8; 2/6 = 1/3; 4/2 = 2/1 = 8/4 = 16/8; 3/4 = 6/8; 10/2 = 5/1

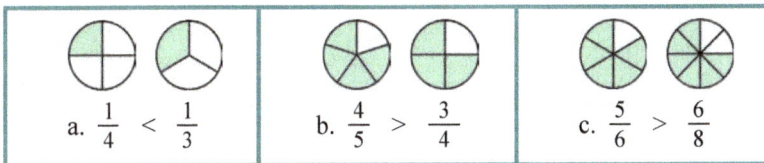

Puzzle Corner: a. 15   b. 13   c. 60   d. 25

Comparing Fractions 1, pp. 191-193

1.

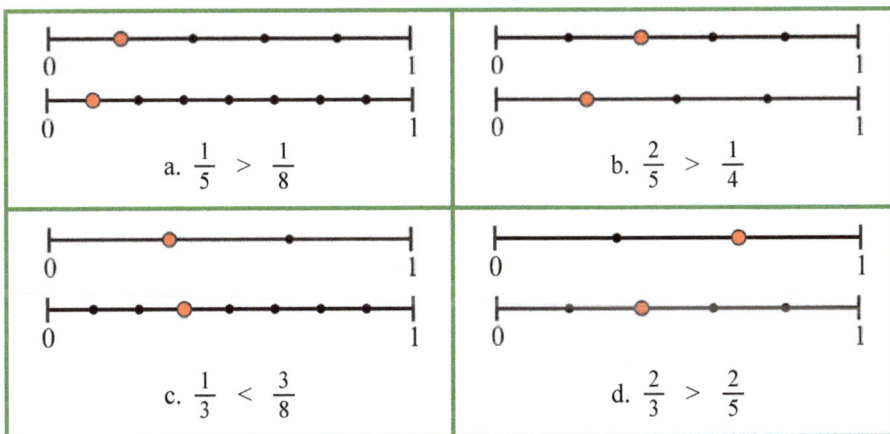

2.

a. $\frac{1}{5} > \frac{1}{8}$	b. $\frac{2}{5} > \frac{1}{4}$
c. $\frac{1}{3} < \frac{3}{8}$	d. $\frac{2}{3} > \frac{2}{5}$

3. a. (cross out)   b. 7/8 < 9/10
   c. 6/8 < 5/6   d. 3/8 > 3/10

4. Hazel is correct. While both fractions are equal to one whole, in this case the wholes are not the same size.

Comparing Fractions 1, cont.

5. a. 2/3 > 1/3   b. 1/5 < 4/5   c. 1/6 < 3/6
   d. (cross out)   e. (cross out)   f. 7/10 > 4/10

6. You can just compare how many parts there are. For example, 5/8 has more eighths than 3/8, so it is the bigger fraction. In other words, just compare the numerators. 7/6 has more sixths than 1/6, so it is bigger.

7. a. $\frac{3}{4} < \frac{5}{4}$ b. $\frac{3}{6} > \frac{1}{6}$ c. $\frac{9}{8} > \frac{8}{8}$ d. $\frac{5}{5} > \frac{2}{5}$

   e. $\frac{13}{10} > \frac{3}{10}$ f. $\frac{3}{3} < \frac{5}{3}$ g. $\frac{3}{6} < \frac{6}{6}$ h. $\frac{5}{2} > \frac{2}{2}$

8. No. The bigger pitcher has more. 1/4 of something that is larger is more than 1/4 of something that is smaller.

9. It is hard to tell who got more pie to eat. You cannot really tell. Notice that the wholes are not the same size.

Comparing Fractions 2, pp. 194-196

1.

a. $\frac{1}{3} < \frac{1}{2}$	b. $\frac{1}{2} > \frac{1}{5}$	
c. $\frac{1}{5} < \frac{1}{4}$	d. $\frac{1}{6} < \frac{1}{5}$	
e. $\frac{1}{6} > \frac{1}{8}$	f. $\frac{1}{2} > \frac{1}{8}$	

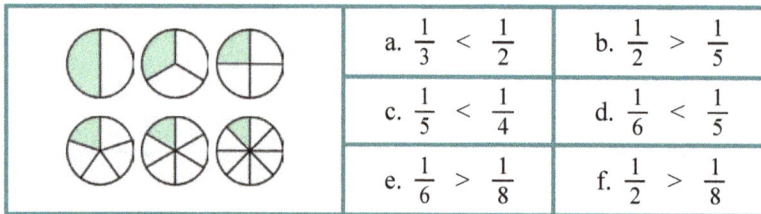

2. a. One eighth (1/8) is the bigger fraction. If you divide a whole into 8 pieces, each piece is bigger than if you divide a whole into 9 pieces.
   b. For unit fractions, the larger the denominator, the smaller the fraction.

3. Since 1/3 is further from zero than 1/4, it is the bigger fraction.

4. The number lines are not of the same length so are representing the fractions incorrectly. When we make the two number lines to have the same length from 0 to 1, we see that 1/5 < 1/4.

You can also use pie pictures:

5. a.

$\frac{2}{2}$

$\frac{2}{4}$

$\frac{2}{5}$

$\frac{2}{10}$

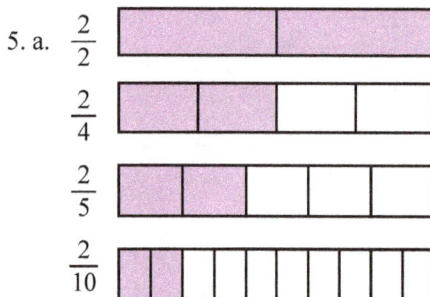

b. The fractions all have the same numerator (number of parts). They each have a different denominator (kind of parts).

c. The size of each fraction, compared to the others, is based on the size of its parts. When a whole is divided into more parts, the parts are smaller.

Comparing Fractions 2, cont.

Page 195

6.

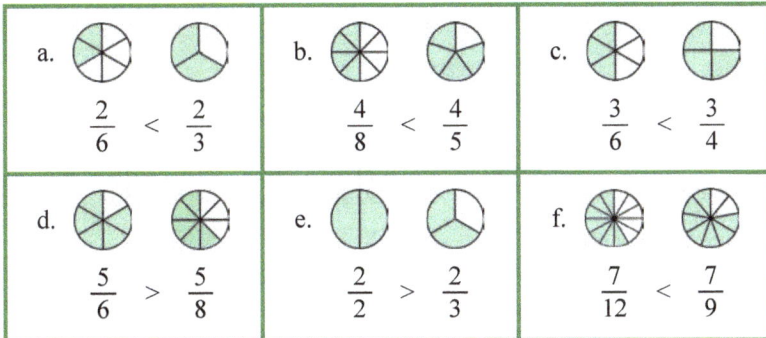

a. $\dfrac{2}{6} < \dfrac{2}{3}$	b. $\dfrac{4}{8} < \dfrac{4}{5}$	c. $\dfrac{3}{6} < \dfrac{3}{4}$
d. $\dfrac{5}{6} > \dfrac{5}{8}$	e. $\dfrac{2}{2} > \dfrac{2}{3}$	f. $\dfrac{7}{12} < \dfrac{7}{9}$

7. You can just check what kind of pieces the two fractions have, and choose the fraction that has bigger pieces. For example, 5/8 has eighths, and eighths are bigger pieces than ninths, so 5/8 is more than 5/9. Eighths are bigger pieces than tenths, so 6/8 is bigger than 6/10. In other words, just compare the denominators, and the fraction with the smaller denominator is the greater fraction.

Page 196

8. a. >   b. <   c. >   d. >   e. =   f. <   g. =   h =

9. a. >   b. >   c. >   d. =   e. <   f. =   g. <   h <

10. Answers will vary. Check the student's answer. For example: Think of the fractions as so many pieces. Perhaps draw a picture. In this case, we can see that 5 thirds is more than 1, and 7 eighths is less than 1, so 5/3 is greater than 7/8.

11. $\dfrac{1}{9} < \dfrac{1}{6} < \dfrac{1}{5} < \dfrac{1}{3}$

Puzzle Corner

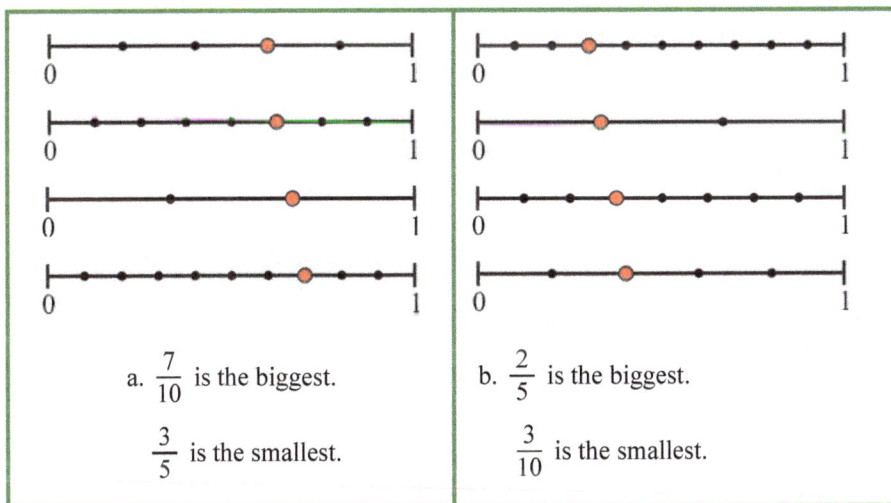

a. $\dfrac{7}{10}$ is the biggest.

$\dfrac{3}{5}$ is the smallest.

b. $\dfrac{2}{5}$ is the biggest.

$\dfrac{3}{10}$ is the smallest.

Comparing Fractions 3, pp. 197-198

Page 197

1. a. (cross out)   b. 2/9 < 2/6   c. (cross out)
   d. 3/10 > 1/4   e. (cross out)   f. 2/7 > 2/9

2. A piece from the red ribbon. That is because 1/4 > 1/5.

Comparing Fractions 3, cont.

3. a. $\dfrac{4}{3} > \dfrac{3}{3}$ b. $\dfrac{6}{7} > \dfrac{6}{9}$ c. $\dfrac{9}{10} > \dfrac{7}{10}$ d. $\dfrac{9}{12} < \dfrac{9}{5}$

  e. $\dfrac{1}{6} < \dfrac{1}{4}$ f. $\dfrac{1}{12} < \dfrac{10}{10}$ g. $\dfrac{3}{8} < \dfrac{3}{6}$ h. $\dfrac{1}{2} < \dfrac{8}{8}$

4. The bar model for 9/10 was shorter than the one for 7/8, so it wasn't an accurate comparison.

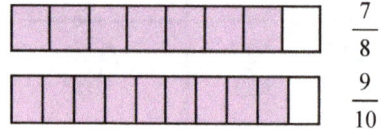

   Now we have two wholes that are the same size. Now we can compare, and see that 9/10 is slightly greater than 7/8. We write: 9/10 > 7/8

5.

a. $\dfrac{5}{8}$ is the biggest.

$\dfrac{2}{4}$ is the smallest.

b. $\dfrac{7}{8}$ is the biggest.

$\dfrac{2}{3}$ is the smallest.

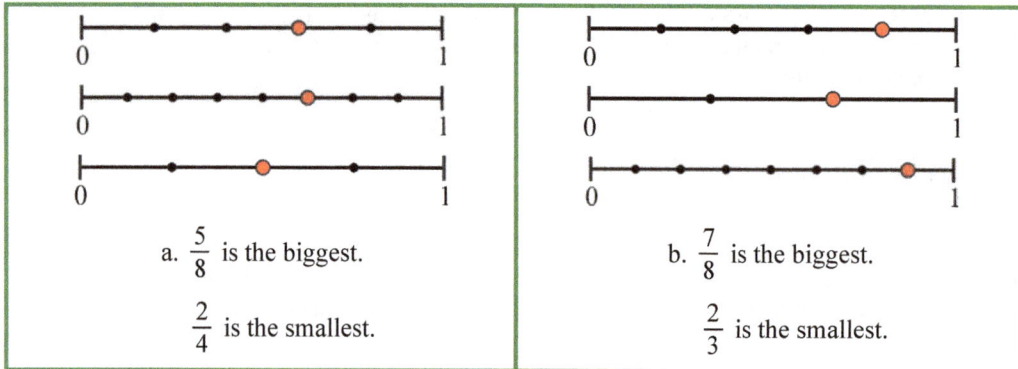

6. One-twelfth of the bigger bar is more to eat than 2/12 of the smaller. However, that does not prove that 1/12 > 2/12 because we were not using wholes of the same size.

7. No, it will not be fair. One-third of the large pizza is a bigger piece to eat than one-third of the small pizza.

Puzzle Corner: $\dfrac{5}{3} < \dfrac{8}{4} < \dfrac{13}{5} < \dfrac{18}{6} < \dfrac{36}{10}$

Mixed Revision Chapter 11, pp. 199-202

1.

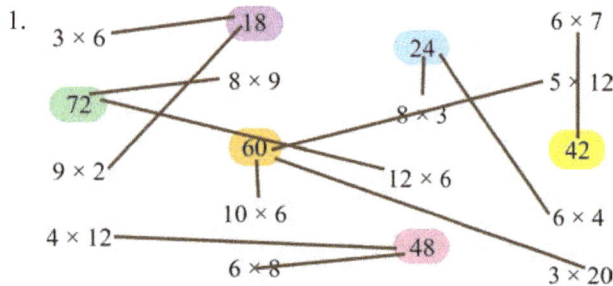

2. a. Each number in the list of 6s will be double the number in the same position in the list of 3s.
   b. The answer will be twice as much: $2 \times 54 = 108$.

3.

a.	b.	c.
$5 \times 6 = 30$	$7 \times 0 = 0$	$1 \times 8 = 8$
$30 \div 5 = 6$	$0 \div 7 = 0$	$8 \div 1 = 8$
$30 \div 6 = 5$	~~$7 \div 0$~~ (not possible)	$8 \div 8 = 1$

Page 199

4. a. 560   b. 900   c. 200
   d. 3000   e. 2000   f. 540

Page 200

5. a. No, she cannot buy them. The price $26.55 is a little more than $25, so, since she wants two sets, the total will
       be more than $50.
   b. The cost for two is $26.55 + $26.55 = $53.10. She needs <u>$3.10 more</u>.

6. a. $3.45   b. $4.15   c. $13.30

7. a. 2 December
   b. 5 January
   c. 13 December
   d. 1 February

8. 28 m × 2 m = 56 m^2

Page 201

9. Drawings will vary. Check the student's drawings. Examples:

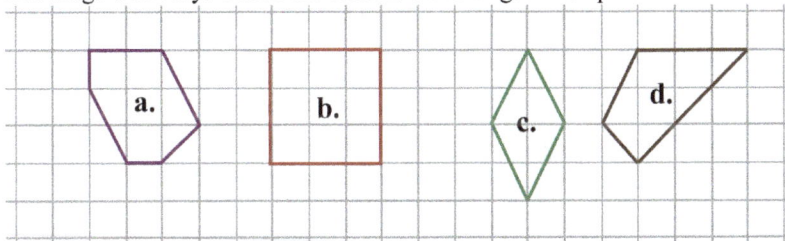

10. 2 × 3 + 4 × 6 = 30 OR 6 × 3 + 4 × 3 = 30 OR 6 × 6 − 2 × 3 = 30

11. 9 m^2

12. a. Estimate: 120 m + 70 m + 120 m + 70 m = 380 m.
    b. The exact perimeter: 376 m

Page 202

13. Total area = 560 m^2
    Area of each part = 80 m^2

14. Perimeter: 6 + 3 + 3 + 4 + 3 + 7 = 26; 26 cm
    Area: 3 × 3 + 7 × 3 = 30; 30 cm^2

15.

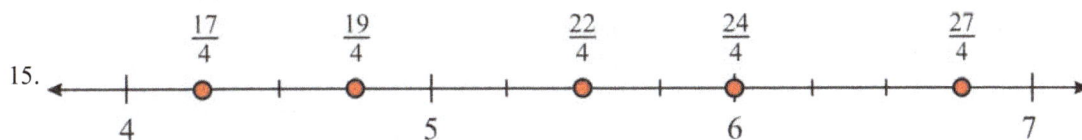

16. Since the two fractions have the same amount of pieces, look at the size of
    the pieces. Eighths are larger pieces than tenths, so 7/8 is greater than 7/10.

17. No. She is not correct. The two wholes are not the same size. If the wholes
    are made the same size, we can easily see that 2/8 < 2/4.

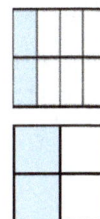

1.

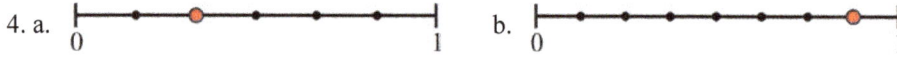

2. a. 3/6   b. 13/8   c. 6/3   d. 3/2

3. a. 1/5   b. 6/8

4. a. 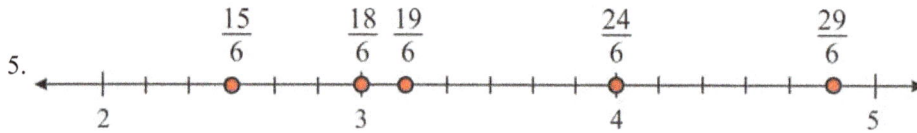 b.

5.

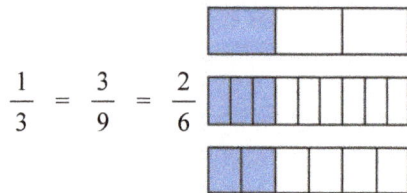

6. Since 30/6 = 5 and 36/6 = 6, 35/6 is between 5 and 6.

Page 204

7. a. 1 = 2/2   b. 2 = 20/10   c. 2 = 8/4
   d. 4 = 20/5   e. 7 = 42/6   f. 4 = 40/10   g. 6 = 12/2   h. 5 = 40/8

8.

a. $\frac{2}{5} = \frac{4}{10}$ b. $\frac{1}{2} = \frac{6}{12}$ c. $\frac{9}{12} = \frac{3}{4}$

9.

$\frac{1}{3} = \frac{3}{9} = \frac{2}{6}$

10. Answers will vary. Students may use pie models, rectangular models,
    number lines, squares, or other shapes to show this. For example:

11. a. 7/10 < 3/4   b. 5/8 < 3/4   c. (cross out)

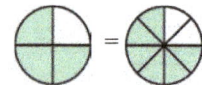

Page 205

12. a. 6/8 < 7/8   b. 1/5 > 1/8   c. 3/8 < 3/5   d. 1/2 = 2/4   e. 24/10 > 15/10

13. Looking at the numerators, we can see that both fractions have 5 pieces. Then considering the denominators,
    or the kind of pieces, halves are greater than sixths, so 5/2 is larger than 5/6.

14. $\frac{2}{9} < \frac{1}{3} < \frac{4}{9} < \frac{3}{6}$

15. a. The bigger can has more paint.
   b. In this case, it does not, because the wholes (paint cans) are not the same size.

Puzzle Corner: The particular squares that are shaded may vary. Check that it is the correct fractional part. For example:

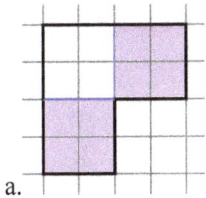

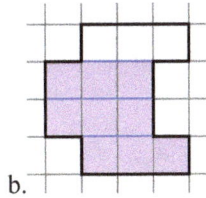

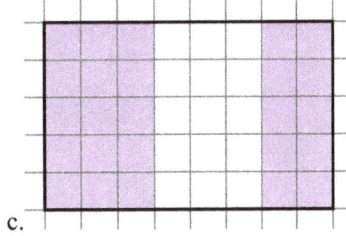

a.

b.

c.

Test Answer Keys

Math Mammoth Grade 3
International Version
Tests Answer Key

Chapter 1 Test

Grading

My suggestion for grading the chapter 1 test is below. The total is 21 points. Divide the student's score by the total of 21 to get a decimal number, and change that decimal to percent to get the student's percentage score.

Question	Max. points	Student score
1	6 points	
2	3 points	
3	3 points	

Question	Max. points	Student score
4	3 points	
5	6 points	
Total	21 points	

1. a. 270; 203   b. 93; 129   c. 47; 871

2. a. 5   b. 287   c. 8

3. a. 6   b. 4   c. 10

4.

$$130 + x = 290 \ (\text{or } x + 130 = 290)$$
$$290 - x = 130 \ (\text{or } 290 - 130 = x)$$
$$x = 160$$

5. a. $238 + 9 + 10 = n$
   Solution: $n = 257$ The new rent will be $257.

   b. $90 + 90 - 60 = h$
   Solution: $h = 120$ He has $120 left.

Chapter 2 Test

Grading

My suggestion for grading the chapter 2 test is below. The total is 24 points. Divide the student's score by the total of 24 to get a decimal number, and change that decimal to percent to get the student's percentage score.

Question	Max. points	Student score
1	4 points	
2	4 points	
3a	1 point	
3b	2 points	
3c	2 points	

Question	Max. points	Student score
4	2 points	
5	2 points	
6	4 points	
7	3 points	
Total	24 points	

1. a. 710   b. 600   c. 820   d. 460

2. a. 247; 247 + 157 = 404   b. 326; 326 + 397 = 723

3. a. Answers may vary, depending on how the student rounds the numbers. This estimate rounds to the nearest ten:
   60 + 490 + 80 + 150 = 780
   b. Exact: 778
   c. Yes, the answer is reasonable, because it is very close to the estimate.

4. Answers may vary. Rounding to the nearest 10:
   $250 - (130 + 20) = 100$. He has approximately $100 left.

5. 189 days are not school days.

6. a. 27   b. 43   c. 310   d. 310

7. $609 - (169 + 145) = 295$. First add: $169 + 145 = 314$. Then subtract: $609 - 314 = 295$.

Chapter 3 Test

Grading

My suggestion for grading the chapter 3 test is below. The total is 28 points. Divide the student's score by the total of 28 to get a decimal number, and change that decimal to percent to get the student's percentage score.

Question	Max. points	Student score
1	12 points	
2	4 points	

Question	Max. points	Student score
3	8 points	
4	4 points	
Total	28 points	

1. a. 6, 5, 0   b. 10, 30, 12   c. 40, 120, 400   d. 9, 0, 11

2. Answers will vary. Check students' answers. For example:

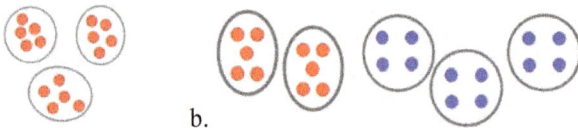

a.   b.

3. a. $3 \times 12 = 36$; The three baskets have 36 apples.   b. $5 \times 4 = 20$; She got 5 groups of 4 sticks.
   c. $4 \times \$2 + 2 \times \$8 = \$24$; The total cost was $24.   d. $6 \times 5 = 30$; It had 5 columns of flowers.

4. a. 20   b. 22   c. 0   d. 11

Chapter 4 Test

Grading

Question 1 is to be timed but it's best if you don't let the student know how long a time they have, to avoid anxiety. Allow 3 minutes of time. The student does not have to finish all the questions; there are more questions than what is needed for a perfect score. The point count depends on the number of correct answers as follows: Divide the number of correct answers by 2, and that is the point count, up to 16 points. In other words, a student that gets 32 or more questions correct will get 16 points (no more).

My suggestion for grading the chapter 4 test is below. The total is 38 points. Divide the student's score by the total of 38 to get a decimal number, and change that decimal to percent to get the student's percentage score.

Question	Max. points	Student score
1	16 points	
2	1 point	
3	2 points	
4	2 points	

Question	Max. points	Student score
5	3 points	
6	8 points	
7	6 points	
Total	38 points	

1. a. 16, 49, 24, 45   b. 25, 32, 35, 48   c. 8, 36, 20, 44   d. 42, 0, 16, 45
   e. 24, 24, 60, 36   f. 32, 9, 96, 30   g. 72, 12, 66, 63   h. 49, 88, 54, 21
   i. 10, 64, 18, 36   j. 36, 15, 20, 48   k. 30, 40, 28, 48   l. 56, 72, 18, 42

2. $9 \times 7 = 63$ or $7 \times 9 = 63$

3. $2 \times 51 = 102$; $2 \times 102 = \underline{204}$

4. The exact way the student breaks the multiplication into two parts will vary. For example:
$17 \times 7 = 10 \times 7 + 7 \times 7 = 70 + 49 = 119$ or $17 \times 7 = 8 \times 7 + 9 \times 7 = 56 + 63 = 119$.

5. a. 45   b. 70   c. 36

6. a. $5 \times \$9 + 5 \times \$5 = \$70$. They could cost $70.
b. $9 \times 6 = 54$. You need nine tables.
c. $7 \times 4 + 4 \times 4 + 12 \times 2 = 68$. They have 68 feet in total.
d. $8 \times \$6 = \$48$. You can buy 8 shirts.

7. a. 4, 9, 6   b. 11, 6, 2   c. 6, 9, 4   d. 8, 2, 4

Chapter 5 Test

Grading

My suggestion for grading the chapter 5 test is below. The total is 22 points. Divide the student's score by the total of 18 to get a decimal number, and change that decimal to percent to get the student's percentage score.

Question	Max. points	Student score
1	6 points	
2	2 points	
3	4 points	
4	4 points	

Question	Max. points	Student score
5	2 points	
6	2 points	
7	2 points	
Total	22 points	

1. a. 1:47; 1:57   b. 5:34; 5:44   c. 3:57; 4:07

2. a. 22 minutes   b. 39 minutes

3. a. 35 minutes   b. 22 minutes

4. a. 2 hours   c. 43 minutes
   b. 20 minutes   d. 34 minutes

5. He was gone for three hours.

6. They returned on 4 October.

7. It was 2:20 PM.

Chapter 6 Test

Grading

My suggestion for grading the chapter 6 test is below. The total is 15 points. Divide the student's score by the total of 15 to get a decimal number, and change that decimal to percent to get the student's percentage score.

Question	Max. points	Student score
1	2 points	
2	3 points	
3	4 points	

Question	Max. points	Student score
4	2 points	
5	4 points	
Total	15 points	

1. a. $25.75   b. $4.60

2. a. $1.20   b. $3.60   c. $4.20

3. a. $9.20   b. $0.80 or 80 cents.

4. Marsha still needs to save $16.85.

5. a. The total cost is $15.50.
   b. His change is $4.50.

Chapter 7 Test

Grading

My suggestion for grading the chapter 7 test is below. The total is 22 points. Divide the student's score by the total of 22 to get a decimal number, and change that decimal to percent to get the student's percentage score.

Question	Max. points	Student score
1	4 points	
2	4 points	
3	4 points	

Question	Max. points	Student score
4	3 points	
5a	4 points	
5b	3 points	
Total	22 points	

1. a. 2689   b. 4070
   c. 5609   d. 3902

2. a. >   b. >
   c. <   d. <

3. a. 700; 8200
   b. 8100; 8100

4. 8085

5. a. $2000 − ($1529 + $325) = $86; Their change is $86.
   b. $2566 − $650 = $1916; The used computer is $1916 cheaper.

Chapter 8 Test

Grading

My suggestion for grading the chapter 8 test is below. The total is 32 points. Divide the student's score by the total of 32 to get a decimal number, and change that decimal to percent to get the student's percentage score.

Question	Max. points	Student score
1	2 points	
2	3 points	
3	12 points	
4	4 points	

Question	Max. points	Student score
5a	2 points	
5b	2 points	
6	5 points	
7	2 points	
Total	32 points	

1.

or

2. $6 \times 8 = 48$
   $8 \times 6 = 48$
   $48 \div 6 = 8$
   $48 \div 8 = 6$

3. a. 8, 4, 8   b. 9, 10, 7   c. 7, 4, 8   d. 0, 12, 1

4. a. 14   b. 7   c. 110   d. 56

5. a. $54 \div 6 = 9$ or $6 \times 9 = 54$. They will make 9 groups.
   b. $(17 + 7) \div 2 = 12$. Each boy got 12 crayons.

6. The student may use a different number for the chick picture to represent.

Chicks Hatched		
Week 8	200	🐤🐤🐤🐤
Week 9	150	🐤🐤🐤
Week 10	225	🐤🐤🐤🐤🐤
Week 11	175	🐤🐤🐤🐤

🐤 = 50 chicks

🐤 = 25 chicks

7. a. 25 more chicks hatched in week 11.
   b. 750 chicks hatched during these four weeks.

Chapter 9 Test

Grading

My suggestion for grading the chapter 9 test is below. The total is 20 points. Divide the student's score by the total of 20 to get a decimal number, and change that decimal to percent to get the student's percentage score.

Question	Max. points	Student score
1	2 points	
2	2 points	
3	10 points	

Question	Max. points	Student score
4	2 points	
5	2 points	
6	2 points	
Total	20 points	

1. a. ——————————————————————————

   b. ——————————————————————————

2.
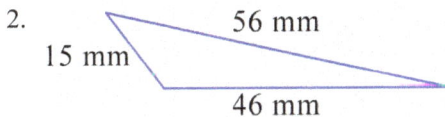
56 mm
15 mm
46 mm

3.

a. Mary's book weighed 350 _g_ .	f. Mum bought 3 _kg_ of bananas.
b. A box of juice had 2 _L_ of juice.	g. Erika weighs 55 _kg_ .
c. The aeroplane was flying 10 000 _metres_ above the ground.	h. A cell phone weighs 120 _g_ .
d. The large tank holds 200 _L_ of water.	i. A housefly measured 7 _mm_ long.
e. Andy and Matt bicycled 10 _km_ to the beach.	j. The shampoo bottle can hold 450 _ml_ of shampoo.

4. Five T-shirts would have a mass of 1 kg.

5. Each container has 4 litres of water.

6. a. 70 ml   b. 250 ml   c. 180 ml

Chapter 10 Test

Grading

My suggestion for grading the chapter 10 test is below. The total is 25 points. Divide the student's score by the total of 25 to get a decimal number, and change that decimal to percent to get the student's percentage score.

Question	Max. points	Student score
1	7 points	
2	2 points	
3	2 points	
4	4 points	

Question	Max. points	Student score
5	2 points	
6	3 points	
7	2 points	
8	3 points	
Total	25 points	

1. a. Shape 2
   b. Shape 3
   c. Shape 5
   d. Shape 1
   e. Shape 4
   f. Shape 6
   g. Shape 7

2. Student drawings will vary. Check the student's drawings. Below are some examples.

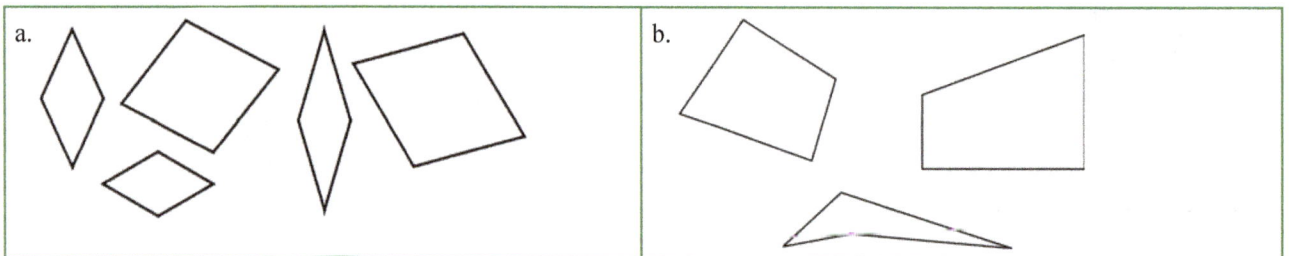

3. Area = 9 square units   Perimeter = 14 units

4. In this problem it is required that the student give the correct *unit*, not just the correct number.
   a. Perimeter = 12 m   Area = 8 m^2 b. Perimeter = 28 cm   Area = 49 cm^2

5. $2 \times 3 + 6 \times 3 = 24$ square units   OR $2 \times 3 + 3 \times 6 = 24$ square units
   OR $3 \times 2 + 6 \times 3 = 24$ square units   OR $3 \times 2 + 3 \times 6 = 24$ square units

6. Divide the shape into two rectangles (which can be done in two different ways).
   Area = 4 m $\times$ 11 m + 4 m $\times$ 8 m = 76 m^2 OR 7 m $\times$ 4 m + 12 m $\times$ 4 m = 76 m^2.
   It is required that the student include <u>square metres</u> with his/her answer (m^2), not just the correct number.

7. The latter pen has a larger perimeter. Its perimeter is 96 m, whereas the perimeter of the first pen is 84 m. The difference is 12 m. It is required that the student include metres with their answer (m), not just the correct number.

8.

$4 \times (2 + 5) = 4 \times 2 + 4 \times 5$

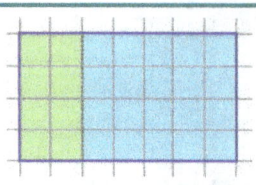

Chapter 10 Test (Version 2)

Grading

My suggestion for grading the chapter 10 test is below. The total is 25 points. Divide the student's score by the total of 25 to get a decimal number, and change that decimal to percent to get the student's percentage score.

Question	Max. points	Student score
1	6 points	
2	1 points	
3	2 points	
4	2 points	
5	4 points	

Question	Max. points	Student score
6	2 points	
7	3 points	
8	2 points	
9	3 points	
Total	25 points	

1. a. Shape 4
   b. Shape 1
   c. Shape 5
   d. Shape 6
   e. Shape 3
   f. Shape 2

2. Student drawings will vary. Check the student's drawing. Below are some examples. The two congruent sides are in the same colour and also marked with single tick marks.

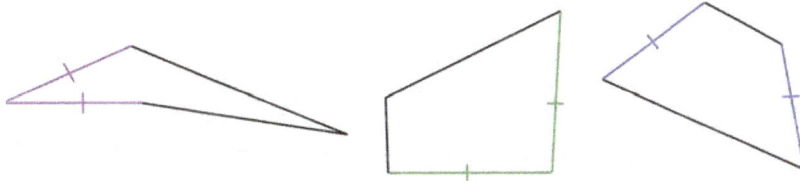

3. Area = 10 square units   Perimeter = 16 units

4. $21 \text{ cm} + x = 54 \text{ cm}$ or $21 \text{ cm} + x + 21 \text{ cm} + x = 108 \text{ cm}$
   Solution: $x = 33$ cm

5. In this problem it is required that the student give the correct *unit*, not just the correct number.
   a. Perimeter = 16 m   Area = 15 m^2 b. Perimeter = 36 m   Area = 81 m^2

6. $2 \times 6 + 3 \times 3 = 21$ square units   OR $6 \times 2 + 3 \times 3 = 21$ square units

7. Divide the shape into two rectangles (which can be done in two different ways).
   Area = 3 m × 12 m + 5 m × 8 m = 76 m^2.   OR   3 m × 7 m + 5 m × 11 m = 76 m^2.
   It is required that the student include square metres with his/her answer (m^2), not just the correct number.

8. Lindsay's garden has a larger perimeter. Its perimeter is 24 m, whereas the perimeter of her neighbour's garden is 22 m. The difference is 2 m. It is required that the student include metres with their answer (m), not just the correct number.

9. $4 \times (3 + 2)$ = 4×3 + 4×2

      area of the   area of the   area of the
   whole rectangle   first part   second part

Chapter 11 Test

Grading

My suggestion for grading the chapter 11 test is below. The total is 27 points. Divide the student's score by the total of 27 to get a decimal number, and change that decimal to percent to get the student's percentage score.

Question	Max. points	Student score
1	3 points	
2	2 points	
3	5 points	
4	4 points	
5	4 points	

Question	Max. points	Student score
6	2 points	
7	3 points	
8	2 points	
9	2 points	
Total	27 points	

1. Student drawings will vary; check the student's drawing. There are multiple ways to divide the shapes. Below are some examples.

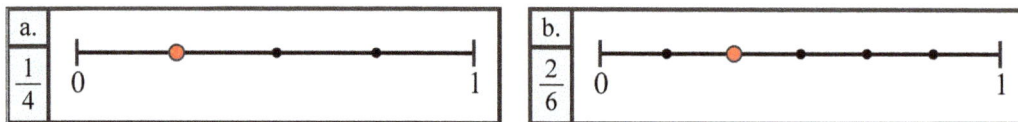

2.

3.

4.

a. $1 = \dfrac{6}{6}$ b. $2 = \dfrac{8}{4}$ c. $4 = \dfrac{32}{8}$ d. $5 = \dfrac{15}{3}$

5. a. <   b. >   c. >   d. >

6. $\dfrac{1}{6} < \dfrac{1}{3} < \dfrac{2}{2}$

7.

a. $\dfrac{2}{5} = \dfrac{4}{10}$ b. $\dfrac{6}{8} = \dfrac{3}{4}$ c. $\dfrac{2}{3} = \dfrac{6}{9}$

8. They eat the same amount of bread, because eating 3 pieces out of 12, and 2 pieces out of 8 signify the fractions 3/12 and 2/8, and they are equivalent fractions (both are in fact equal to 1/4).

9. It is wrong because the two wholes that we take the fractions from are not the same size. You cannot compare fractions unless the wholes they refer to are the same size.

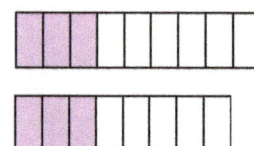

$\dfrac{3}{9} = \dfrac{3}{8}$

138

End-of-Year Test Grade 3
Grading

Instructions to the teacher: My suggestion for grading is below. The total is 219 points. A score of 175 points is 80%.

Grading on question 1 (the multiplication tables grid): There are 144 empty squares to fill in the table, and the completed table is worth 14 points. Count how many of the answers the student gets right, divide that by 10, and round to the nearest whole point. For example: a student gets 24 right. 24/10 = 2.4, which rounded becomes 2 points. Or, a student gets 85 right. 85/10 = 8.5, which rounds to 9 points.

Grading on question 2: Each question is worth 1/2 point.

Question	Max. points	Student score
Multiplication Tables and Basic Division Facts		
1	14 points	
2	8 points	
3	8 points	
subtotal		/ 30
Addition and Subtraction		
4	6 points	
5	6 points	
6	3 points	
7	2 points	
8	3 points	
subtotal		/ 20
Regrouping and Rounding		
9	3 points	
10	2 points	
11	4 points	
12	3 points	
13	4 points	
14	3 points	
subtotal		/ 19
Multiplication and Related Concepts		
15	1 point	
16	1 point	
17	3 points	
18	3 points	
19	3 points	
20a	2 points	
20b	2 points	
20c	2 points	
20d	2 points	
subtotal		/ 19

Question	Max. points	Student score
Time		
21	6 points	
22	2 points	
23	4 points	
23	4 points	
subtotal		/ 16
Graphs		
25	4 points	
26	3 points	
subtotal		/ 7
Money		
27	4 points	
28	3 points	
29	3 points	
subtotal		/ 10
Four-Digit Numbers		
30	2 points	
31	2 points	
32	5 points	
33	4 points	
34	4 points	
subtotal		/ 17
Division and Related Concepts		
35	2 points	
36	9 points	
37	6 points	
38	6 points	
subtotal		/ 23

Question	Max. points	Student score
Measuring		
39	2 point	
40	1 point	
41	1 point	
42	1 point	
43	6 points	
subtotal		/ 11
Geometry		
44	6 points	
45	3 points	
46	2 points	
47	3 points	
48	2 points	
49	2 points	
50	2 points	
subtotal		/ 20

Question	Max. points	Student score
Fractions		
51	5 points	
52	5 points	
53	4 points	
54	3 points	
55	2 points	
56	5 points	
57	3 points	
subtotal		/ 27
TOTAL		/ 219

Grade 3 End-of-Year Test Answers

1.

×	1	2	3	4	5	6	7	8	9	10	11	12
1	1	2	3	4	5	6	7	8	9	10	11	12
2	2	4	6	8	10	12	14	16	18	20	22	24
3	3	6	9	12	15	18	21	24	27	30	33	36
4	4	8	12	16	20	24	28	32	36	40	44	48
5	5	10	15	20	25	30	35	40	45	50	55	60
6	6	12	18	24	30	36	42	48	54	60	66	72
7	7	14	21	28	35	42	49	56	63	70	77	84
8	8	16	24	32	40	48	56	64	72	80	88	96
9	9	18	27	36	45	54	63	72	81	90	99	108
10	10	20	30	40	50	60	70	80	90	100	110	120
11	11	22	33	44	55	66	77	88	99	110	121	132
12	12	24	36	48	60	72	84	96	108	120	132	144

2. a. 14, 24, 25, 36   b. 28, 40, 27, 35   c. 9, 16, 49, 32   d. 56, 30, 48, 54

3. a. 7, 5, 8, 7   b. 8, 5, 11, 7   c. 9, 7, 4, 9   d. 10, 8, 3, 3

4. a. 310, 149   b. 620, 344   c. 148, 80

5. a. 33, 5   b. 643, 45   c. 15, 378

6.

total 900	
440	460

$460 + 440 = 900$
$900 - 460 = 440$
$900 - 440 = 460$

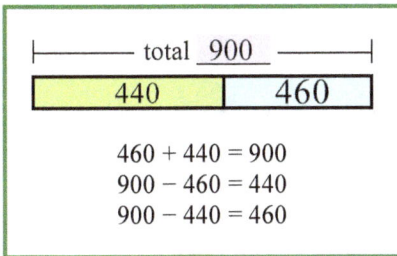

7. 160 kilometres. Note that the half-way point is at 150 km. They stopped at 140 km (10 km before 150 km).

8. Equation: $400 + $400 − $600 = $200 (or 2 × $400 − $600 = $200)
   Solution: He took $200 off the price.

9. a. 90   b. 610   c. 460

10. Round $38 to $40. 4 × $40 = $160 and 5 × $40 = $200. He can buy it after 5 weeks.

11. a. △ is 294. Solve by subtracting 708 − 414.   b. △ is 824. Solve by adding 485 + 339.

12. a. $545 + $52 = $310 + m
    b. Estimates may vary. $550 + $50 = $600; $600 − $310 = $290. He needs about $290 more.
    c. $545 + $52 = $597; $597 − $310 = $287. He needs $287 more.

13. a. 579. To check, add 579 + 383 = 962 using the grid.   b. 157. To check, add 157 + 549 = 703 using the grid.

14. a. Estimate, rounding to the nearest ten:
       80 + 540 + 150 + 10 = 780
    b. By estimating with easier-to-add numbers, he can see that
       something is wrong with his calculation.
    c. The 8 needs moved to the ones' column.

```
  1 2
      8 2
    5 3 9
    1 5 4
  +     8
  -------
    7 8 3
```

15.

16. 75

17. a. 240   b. 490   c. 300

18. a. 7 × 4 = 28 legs
    b. 5 × 2 = 10 legs
    c. 8 × 4 + 6 × 2 = 44 legs

19. a. 48   b. 20   c. 41

20. a. 3 × 12 = 36. She needs 36 rolls.
    b. 8 × 4 = 32. You need 8 tables.
    c. 3 × $11 + 3 × $8 = $33 + $24 = $57. It would cost $57.
    d. 7 × 4 = 28. She will need 7 bags.

21.

	a. 10:51	b. 5:38	c. 3:57
10 min. later	11:01	5:48	4:07

22. a. 19 minutes   b. 31 minutes

23. a. 23 minutes   b. 33 minutes

24. a. She watched for 17 minutes
    b. It should go in at 5:45 PM.

25. a. 40 hours
    b. 10 hours
    c. 10 hours
    d. 45 hours

26. Three hours is a good number to have each tennis
    ball represent, since each player's practice hours
    is divisible by 3. Two could also be used, with
    half of a ball drawn to represent the odd hours.

Tennis Practice	
Ava	🎾 🎾
Juan	🎾
Greg	🎾 🎾 🎾
Adelaide	🎾 🎾 🎾 🎾

🎾 = 3 hours

27. a. Total: $18.60 Change: $1.40
    b. Total: $29 Change: $1

28. a. $49.54 which is rounded to $49.55   b. $26.40   c. $38.85

29. His change is $26.45.

30. a. 5205   b. 2094
    c. 7300   d. 8002

31. a. 700   b. 2000

32. a. >   b. <   c. <   d. >   e. >

33. a. 5700; 8600   b. 1200; 7800

34. a. 5261; 5261 + 2888 = 8149
    b. 2687; 2687 + 3749 = 6436

35.

$3 \times 6 = 18$ $18 \div 3 = 6$

$6 \times 3 = 18$ $18 \div 6 = 3$

36. a. 10   b. 8   c. 8
    d. 9   e. 5   f. 40
    g. 6   h. 108   i. 8

37. a. 17, not possible   b. 1, not possible   c. 1, 0

38. a. $16 + $14 ÷ 3 = $10. Each child paid $10.00.
    b. 6 × 10 + (10 − 1) = 69 (or 7 × 10 − 1 = 69). There are 69 passengers.
    c. 24 ÷ 6 = 4. They used 4 containers.

39. If you have printed from the digital version at 100% scaling, or you are using the printed book, the answers below
    should be accurate. If you printed from the digital version with "shrink to fit" or "scale to fit" or similar setting, the
    answers below do not match the questions. Check the student's answers.

    a. ─────────────────────────────

    b. ───────────────────────

40. mm   cm   m   km

41. 355 g (grams)

42. There are 12 litres of water.

43. a. km   b. cm
    c. g   d. kg
    e. ml   f. ml

44. a. Shape 2
    b. Shape 3
    c. Shape 1
    d. Shape 4
    e. Shape 6
    f. Shape 5

45. a. Drawings will vary. Check the student's drawing. For example: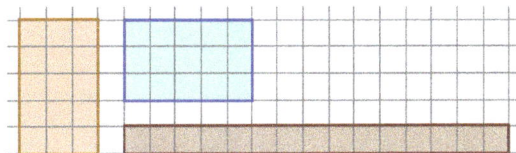

b. Drawings will vary. It should be a rectangle that is not a square. For example: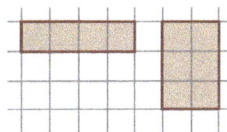

c. Drawings will vary. Check the student's drawing. For example:

46. Perimeter 22 units; area 24 square units or squares.
Note that the student should also give the "units" and "square units" or "squares", not just a plain number.

47. a. Part 1: 108 m^2 Part 2: 270 m^2 b. 96 m
Note that the student should also give the units "m^2" and "m" in his/her answer, not just plain numbers.

48. 9 cm

49. a. The sides of the rectangle could be 5 and 3, or 15 and 1.
Some examples are shown on the right.

b. The sides of the rectangle could be 1 and 4, or 2 and 3.
See the image on the right.

50. $4 \times (2+5) = 4 \times 2 + 4 \times 5 = 28$ squares (or square units)

51. a. $\dfrac{3}{8}$ b. $\dfrac{5}{6}$ c. $\dfrac{8}{3}$ d. $\dfrac{2}{3}$ e. $\dfrac{9}{10}$

52.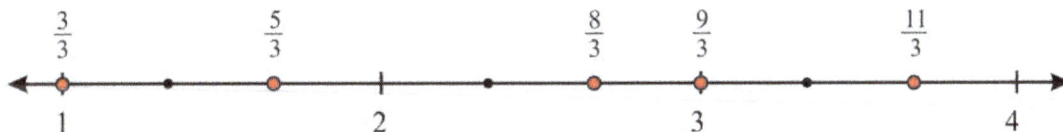

53.

a. $\dfrac{6}{4}$ No	b. $\dfrac{8}{8} = 1$	c. $\dfrac{8}{2} = 4$	d. $\dfrac{2}{8}$ No	e. $\dfrac{13}{3}$ No	f. $\dfrac{24}{4} = 6$	g. $\dfrac{27}{3} = 9$	h. $\dfrac{20}{6}$ No

54.

a. $\dfrac{3}{4} = \dfrac{9}{12}$ b. $\dfrac{10}{12} = \dfrac{5}{6}$ c. $\dfrac{2}{3} = \dfrac{4}{6}$

55.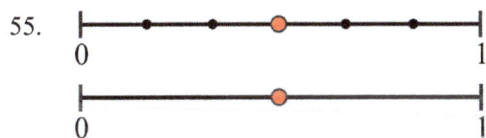

56. a. <   b. <   c. =   d. >   e. =

57. a. Cannot make a valid comparison.   b. =   c. The fractions are 3/8 and 1/3, however you cannot make
a valid comparison, because the shapes are not the same size.

Cumulative Revisions
Answer Keys

Cumulative Revisions Answer Key, Grade 3

Cumulative Revision: Chapters 1 - 2

1. a. 82   b. 83   c. 95
   d. 66   e. 83   f. 35

2. a. 600   b. 810   c. 910
   d. 160   e. 550   f. 240

3. a. 510   b. 760   c. 900   d. 10

4.

a. rent, $256, and groceries, $387	b. an adult's ticket, $58, and child's ticket, $38
rent about $260	adult's ticket about $60
groceries about $390	child's ticket about $40
total about $650	total cost about $100

5.

a. $35 - 14 - 7 + 3 = 17$	d. $(250 - 20) + (80 - 30) = 280$
b. $35 - (14 - 7) + 3 = 31$	e. $250 - (20 + 80 - 30) = 180$
c. $35 - (14 - 7 + 3) = 25$	f. $250 - 20 + (80 - 30) = 280$

6.

+	40	43	46	49	52
32	72	75	78	81	84
34	74	77	80	83	86
36	76	79	82	85	88
38	78	81	84	87	90

7.

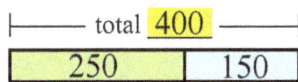
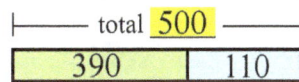

a. $250 + \underline{150} = 400$	b. $390 + \underline{110} = \underline{500}$
$400 - \underline{250} = \underline{150}$	$\underline{500} - \underline{110} = 390$

8. The equations will vary; check the student's equation.

a. Equation: $14 + 13 + c = 30$
Solution: $c = 3$
She needs 3 more cups.

b. Equation: $150 - 60 + 40 = m$ or $m = 150 - 60 + 40$
Solution: $m = 130$
He has $130 now.

Cumulative Revision: Chapters 1 - 3

1.

a.
```
|——— total 698 ———|
| 349 | 196 | 153 |
```
$x + 196 + 153 = 698$

$x = 349$

2. a. 589   b. 316
   c. 258   d. 143

3. Note that one bicycle costs a little less than $100. This means that yes, you can buy three bicycles for $300.
   $96 + 96 + 96 = 288$ and $288 + 12 = 300$. You will have $12 left over. Or, you can notice that one bicycle costs exactly $4 less than $100, which means three bicycles will cost $12 less than $300.

4. a. 9, 0   b. 10, 15   c. 90, 800   d. 0, 22

5. a. <   b. =   c. <

6.

a. $644 - 8 = 636$	b. $277 - 9 = 268$	c. $683 - 8 = 675$
$233 - 7 = 226$	$191 - 5 = 186$	$842 - 7 = 835$

7. a. Estimate: $190 + 380 + 30 + 80 = 680$
   b. She could round the numbers and make an estimate to see if it was reasonably close to her answer.
   c. The 29 needs to be aligned to the right.

8. The equations will vary; check the student's equation.

a. Equation: $480 + (480 - 35) = c$
   Solution: $c = 925$
   The total cost was $925.

b. Equation: $16 + 16 + 40 = m$
   Solution: $m = 72$
   She has $72 now.

```
    2 2
    1 8 6
    3 7 7
      2 9
+     8 2
  ———————
    6 7 4
```

Cumulative Revision: Chapters 1 - 4

1. a. $13 - 9 = 4$ b. $250 - 50 = 200$
   c. 16   d. 140

2. a. $150 + 150 + 150 + 150 = 600$ b. $50 + 50 + 50 = 150$

3. a. 4, 8   b. 0, 9   c. 6, 9   d. 9, 9

4.

a. $8 \times 10 - 2 + 5 = 83$	b. $6 + 7 \times (4 - 2) = 20$
c. $3 \times 4 - 2 \times 3 = 6$	d. $2 \times (4 + 4) \times 2 = 32$

5. Equations will vary; check the student's equation. This time, the prompt is not asking to use an unknown, so using an unknown is optional.

   a. $3 \times 5 = 15$ or $L = 3 \times 5$ or $3 \times 5 = L$, etc. The total length is 15 metres.
   b. $6 + 6 + 6 + 6 + 22 + 22 = 68$ or $4 \times 6 + 2 \times 22 = 68$ or $C = 4 \times 6 + 2 \times 22$ etc. The total cost was $68.

6.

a. $564 - 5 = 559$	b. $888 + 12 = 900$
$564 - 10 = 554$	$886 + 14 = 900$
$564 - 15 = 549$	$884 + 16 = 900$
$564 - 20 = 544$	$882 + 18 = 900$
$564 - 25 = 539$	$880 + 20 = 900$
$564 - 30 = 534$	$878 + 22 = 900$

7. Student illustrations will vary; check the student's illustration. For example, it could be a number line illustration:

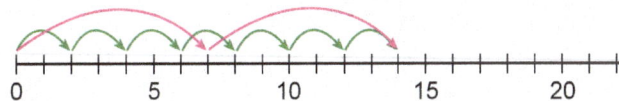

   Or, it could show two sets of 7 objects and seven sets of 2 objects.

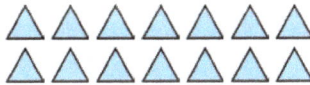

   Or, it could be a 7-by-2 array:

8. It has 4 columns.

9. a. $270 + 270 + 270 = 810$. They cost about $810.
   b. The exact cost is $801. Yes, this answer is reasonable.

Puzzle Corner:
$69 - 21 = 48$

Cumulative Revision: Chapters 1 - 5

1. a. 32   b. 28
   c. 56   d. 37   e. 36

2. a. 577   b. 485

3. Jay forgot to calculate two cupcakes for each of the 11 ladies.
   The correct solution: $30 - 1 - 2 - (11 \times 2) = 5$. There are 5 cupcakes left.

4. 35 minutes

5. a. 22 to 7   b. 4 to 4
   c. 12 past 2   d. 17 to 8

6. a. 35   b. 30   c. 88

7. a. Estimate: $40 + 30 + 30 + 20 + 40 + 30 = 190$. The store sold about 190 bottles of milk.
   b. There were 197 bottles of milk sold in total.
   c. Yes. The exact total is reasonably close to the estimate.

8. a. $78 + s = \$200$ *or* $\$200 - s = \$78$. She spent \$122.
   b. $9 \times 12 - 99 = m$. There were 9 marbles lost.

Cumulative Revision: Chapters 1 - 6

1. a. Estimates will vary. For example: about $160 + 190 + 210 = \underline{560 \text{ km}}$.
   b. They travelled a total of 554 km.

2. a. 390   b. 990   c. 0   d. 350

3. a. 62   b. 61   c. 64
   d. 56   e. 53   f. 50

4. a. 4:36   b. 11:10   c. 11:58   d. 12:43

5. 150 minutes

6. a. 60   b. 36   c. 50

7. Student images will vary; check the student's image. One example is shown on the right. As a single multiplication, $2 \times 4 + 2 \times 7$ is equal to 2×11.

8. It ended at 8:25 PM.

9.

from	8 AM	7 AM	9 AM	11 AM	10 AM
to	12 noon	1 PM	4 PM	11 PM	7 PM
hours	4	6	7	12	9

10.

a.	b.
+ $0.50 + $12.00	+ $1.55 + $10.00
$7.50 $8.00 $20.00	$38.45 $40.00 $50.00
The change is $12.50.	The change is $11.55.

150

Cumulative Revision: Chapters 1 - 7

1. a. 739   b. 994

2. The pattern is to add 11 each time: 420, 431, 442, 453, 464, <u>475</u>, <u>486</u>, <u>497</u>, <u>508</u>, <u>519</u>, <u>530</u>

3. a. 3, 8   b. 10, 5   c. 190, 300

4. $2 \times 7 = 14$; $7 \times 2 = 14$

5. a. 8 hours   b. 25 minutes   c. 35 minutes

6.

a. 4 : 38	b. 3 : 32	c. 2 : 59
22 to 5	28 to 4	1 to 3

7. a. <   b. <
   c. >   d. >

8. a. The total cost was $72.50.
   b. Her change was $27.50.

9. a. $3.30, $4.15   b. $2.70, $6.30   c. $25.30, $15.90

10. a. 12, 9   b. 4, 11   c. 6, 6   d. 12, 6

Cumulative Revision: Chapters 1 - 8

1.

total **910**	total **205**
780 | 130	140 | 65
a. 780 + 130 = 910 910 − 780 = 130 910 − 130 = 780	b. 140 + 65 = 205 205 − 65 = 140 205 − 140 = 65

2.

a. $5 \times 5 = 25$ $12 \times 12 = 144$ $7 \times 5 = 35$	b. $2 \times 11 = 22$ $10 \times 56 = 560$ $4 \times 9 = 36$	c. $7 \times 8 = 56$ $8 \times 12 = 96$ $6 \times 7 = 42$

3. Since she started on 3 November, she will finish on 23 November. Notice that from 3 November to 23 November is 21 days, or three weeks. You will include both 3 November and 23 November in this count of 21 days.

4. It ends at 11:45.

5. a. 155   b. 84   c. 29

6. Equations will vary; check the student's equation. For example: $m - \$39 - \$6 = \$25$. $m = \$70$. She had <u>$70</u> at first.

7. a. 6:45   b. 7:05
   c. 11:45   d. 3:15

8. $2.68 + $2.99 + $2.95 = <u>$8.62</u>

9. a. 5, 9   b. 6, 8   c. 36, 25   d. 12, 9

10. a. △ = 600   b. △ = 500   c. △ = 4600

11. a. 4607   b. 4685

Cumulative Revision: Chapters 1 - 9

1. a. 338. Solve by subtracting $349 - 11 = \underline{338}$.
   b. 210. Solve by subtracting $530 - 320 = \underline{210}$.
   c. 661. Solve by adding $161 + 500 = \underline{661}$.

2. a. See the table on the right.
   b. The difference is about $100.
   c. It costs about $40 more.

	Price	Rounded price
Bob's TV Store	$525	$530
The Nerdy Store	$564	$560
Home Express	$632	$630
Lion Appliances	$599	$600

3. a. 140, 132   b. 144, 121   c. 16, 120

4. a.

$1 \times 9 = 9$	$7 \times 9 = 63$
$2 \times 9 = 18$	$8 \times 9 = 72$
$3 \times 9 = 27$	$9 \times 9 = 81$
$4 \times 9 = 36$	$10 \times 9 = 90$
$5 \times 9 = 45$	$11 \times 9 = 99$
$6 \times 9 = 54$	$12 \times 9 = 108$

   b. For the numbers 1-10, when you multiply by 9, the tens digits go up from 0 to 9 and the ones digits go down from 9 to 0. Also, for any multiple of 9, the sum of its digits will equal 9 (or sometimes the sum of the digits of the number that is the sum of its digits).

5. As usual, student equations may vary. Check the student's equations.
   a. $\$24 \div 4 + \$18 = c$; $c = \$24$. She spent $24 in total.
   b. $(16 + 8) \div 4 = b$; $b = 6$. She needs 6 bags.
   c. $6 \times c = 54$ or $54 \div 6 = c$; $c = 9$. There are 9 chairs in each row.

6. $21 + 14 + 19 = 54$ days

7.

a. $32 \div 8 = 4$ $4 \times 8 = 32$
b. $63 \div 7 = 9$ $9 \times 7 = 63$
c. $35 \div 7 = 5$ $5 \times 7 = 35$

8.

9. In each multiplication, the way it is broken down into two parts will vary. Two ways are given below but there are more. Check the student's answers.

a. 8×15	b. 6×21
$4 \times 15 + 4 \times 15$	$3 \times 21 + 3 \times 21$
$60 + 60 = 120$	$63 + 63 = 126$
or $8 \times 10 + 8 \times 5$	or $6 \times 10 + 6 \times 11$
$80 + 40 = 120$	$60 + 66 = 126$
c. 14×6	d. 4×23
$10 \times 6 + 4 \times 6$	$2 \times 23 + 2 \times 23$
$60 + 24 = 84$	$46 + 46 = 92$
or $7 \times 6 + 7 \times 6$	or $4 \times 20 + 4 \times 3$
$42 + 42 = 84$	$80 + 12 = 92$

10. a. 5, 9 b. 35, 42 c. 8, 9 d. 6, 12

11. a. The student's bar graph may vary. Please check the student's work. Example:

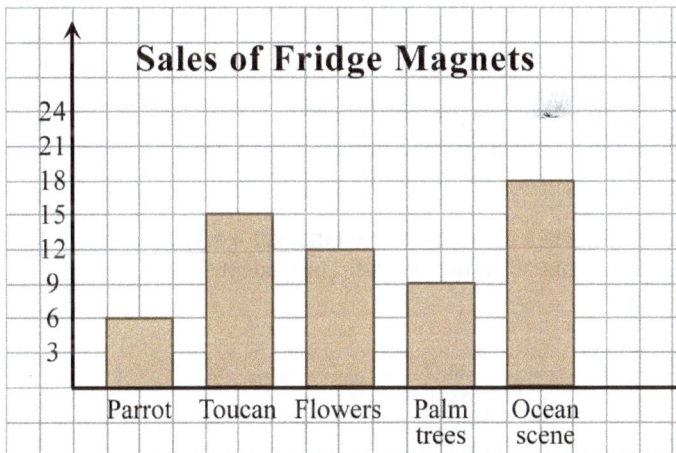

Sales of Fridge Magnets

	24	21	18	15	12	9	6	3
Parrot							6	
Toucan				15				
Flowers					12			
Palm trees						9		
Ocean scene		18						

b. She sold 3 more ocean scene magnets.

c. She sold 21 bird magnets in total, and 39 of all the other kinds, which means she sold <u>18 more</u> of the non-bird magnets.

153

Cumulative Revision: Chapters 1 - 10

1. a. 90   b. 81   c. 44
   d. 76   e. 45   f. 37

2. a. 2   b. 5   c. 9

3.

a.	b.	c.
$7 \times 2 = 14$	$5 \times 7 = 35$	$7 \times 6 = 42$
$2 \times 7 = 14$	$7 \times 5 = 35$	$6 \times 7 = 42$
$14 \div 2 = 7$	$35 \div 5 = 7$	$42 \div 7 = 6$
$14 \div 7 = 2$	$35 \div 7 = 5$	$42 \div 6 = 7$

4. a. 1:15   b. 5:15   c. 7:45   d. 11:45

5. a. 26 minutes   b. 28 minutes   c. 46 minutes

6. a. grams (g)
   b. kilograms (kg)
   c. centimetres (cm)
   d. metres (m) or metres and centimetres

7. a. 2 kg 200 g   b. 3 kg 0 g   c. 0 kg 800 g

8. a. 610   b. 720   c. 300   d. 240

9. a. $49 \div 7 = 7$ *or* $7 \times 7 = 49$. There are 7 rows of cars.
   b. $12 \times (5 + 4 + 2) = 132$. She will have 132 cookies.

10. $5690 < 6050 < 6055 < 6505 < 6553$

11. Rectangles of 3×4, with a perimeter of 14 units, or 2×6, with a perimeter of 16 units will work.
    1×12 would also be 12 square units, though it wouldn't fit in this particular grid. The student should include the word "units" in their perimeter length.

12.

$3 \times (1 + 5)$	=	3×1	+	3×5
area of the whole rectangle		area of the first part		area of the second part

13. a. No, that wasn't correct. Matt's total was $7.25 + $16.90 + $4.60 = $28.75. His change should have been $1.25.

    b. Her total was $2.87 + $2.87 + $2.87 + $2.87 = $11.48 which rounds to $11.50. Her change was $8.50.

Cumulative Revision: Chapters 1 - 11

1. 30 works well for each tree picture to represent, since each number is a multiple of 30. Ten or fifteen could be used, but would require drawing many more trees.

2.

a.	b.	c.	d.
$32 \div 4 = 8$	$12 \times 11 = 132$	$9 = 63 \div 7$	$64 \div 8 = 8$
$0 \div 5 = 0$	$8 \times 0 = 0$	$3 = 18 \div 6$	$45 \div 9 = 5$
$54 \div 6 = 9$	$4 \times 9 = 36$	$1 = 11 \div 11$	$72 \div 9 = 8$

3. The coin was a 1-dollar coin.

4. Answers will vary. Check the student's word problem. For example: You are sharing 24 cookies evenly between 6 people. How many cookies does each person get? Answer: 4.

Silent Creek	🌲🌲🌲🌲
Pine Valley	🌲🌲🌲
Riverside	🌲🌲🌲🌲🌲
Highland	🌲🌲

🌲 = 30 trees

5.

a.	b.	c.
1240 + 50 = <u>1290</u> 8280 − 50 = <u>8230</u>	1090 + 60 = <u>1150</u> 9060 + 40 = <u>9100</u>	3140 − 20 = <u>3120</u> 7780 − 80 = <u>7700</u>

6. 2 kg 300 g *or* 2300 g

7. 4 litres

8. Drawings will vary. Check the student's drawings. For example:

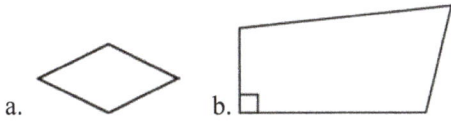

a. b.

9. a. 6 m^2 (or sq m)   b. 10 m

10. a. 5 m × 6 m = 30 m^2 b. 90 m^2 c. 42 m

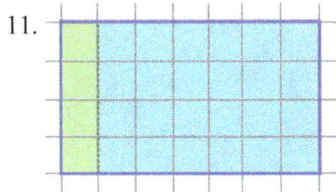

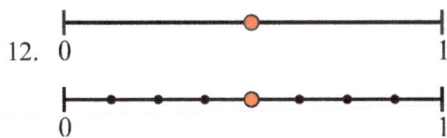

11.

12. 0   1

0   1

13. a. false; 7/1 = 7 *or* 7/7 (or 1/1) = 1   b. true   c. true   d. false; 1/2 = 4/8 *or* 2/3 = 4/6   e. false; 5 = 20/4 *or* 4 = 16/4

14. 3 = 9/3. See the image on the right.

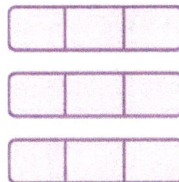

15. 8/12 = 4/6 = 2/3

16. a. >   b. <   c. >   d. >

17. Seven thirds is this much: . It is more than two full pies, or more than two wholes.

   In contrast, seven eighths is less than one whole:

   So, clearly 7/8 is much less than 7/3.

18. No. The shapes are not the same.

19. a. 1795   b. 7593
    c. 5121   d. 2489

20. If the lesson was printed from the digital version with a "Shrink to fit" or similar printer setting (and not scaled at 100%), the student measurements will differ from the ones below. Please check the student's answer.

   The perimeter is 95 mm + 104 mm + 38 mm + 32 mm = <u>269 mm</u>